An Unavoidable Evil

Siege Warfare in the Age of Napoleon

Edited by Zack White

Helion & Company

This book is dedicated to all those across history, whether soldiers or civilians, who have suffered as a result of sieges.

Helion & Company Limited
Unit 8 Amherst Business Centre
Budbrooke Road
Warwick
CV34 5WE
England
Tel. 01926 499619
Email: info@helion.co.uk
Website: www.helion.co.uk
X (formerly Twitter): @Helionbooks
Facebook: @HelionBooks
Visit our blog at http://blog.helion.co.uk/

Published by Helion & Company 2025
Designed and typeset by Mach 3 Solutions (www.mach3solutions.co.uk)
Cover designed by Paul Hewitt, Battlefield Design (www.battlefield-design.co.uk)

Text © Individual contributors 2025
Illustrations © as individually credited
Maps by George Anderson © Helion & Co. 2025
Cover: Siége de Danzic by Johan Lorenz Rugendas (1820). (Anne S.K. Brown Military Collection)

Every reasonable effort has been made to trace copyright holders and to obtain their permission for the use of copyright material. The author and publisher apologise for any errors or omissions in this work, and would be grateful if notified of any corrections that should be incorporated in future reprints or editions of this book.

ISBN 978-1-804513-45-3

British Library Cataloguing-in-Publication Data.
A catalogue record for this book is available from the British Library.

All rights reserved. No part of this publication may be reproduced, stored in a retrieval system, or transmitted, in any form, or by any means, electronic, mechanical, photocopying, recording or otherwise, without the express written consent of Helion & Company Limited.

For details of other military history titles published by Helion & Company Limited, contact the above address, or visit our website: http://www.helion.co.uk

We always welcome receiving book proposals from prospective authors.

Introduction
The Neglected Yet Unavoidable Evil of Siege Warfare in the Age of Napoleon

Zack White

'A conquest of this description is always to be regarded as an evil … we undertake no sieges but such as are positively unavoidable'.[1] These comments on siege warfare by Carl von Clausewitz speak volumes about the awkward placement that sieges had in military thinking in the wake of the Napoleonic Wars. The *Ancien Regime* style of warfare, with its focus upon the capture of border fortifications, had given way in the period 1792–1815 to a prioritisation of agility and the decisive battle.[2] Clausewitz's argument was that sieges should only be undertaken if there either was no other option, or if taking the fortress was the key objective of the campaign. In the latter case, they should not be 'the means but the end itself'.[3] Clearly, therefore, sieges were not considered redundant by the close of the Napoleonic era, yet the necessity of embarking upon them was not to be taken for granted.

With the position of this aspect of warfare being somewhat nebulous, it is unsurprising that the Napoleonic penchant for pitched battles has been reflected in the historiography of the period. Titles undertaking wider-ranging analysis of warfare during the period have offered little, if any, assessment of sieges. Michael Glover, for example, took a largely narrative approach to discussing Wellington's capabilities as a siege commander in his study of the Duke's general-ship, *Wellington as Military Commander*, and even then limited his considerations to the 1812 sieges of Badajoz and Burgos.[4] Although Gunther Rothenberg included an appendix of select sieges in his 1978 study, *The Art of Warfare in the Age of Napoleon*, he devoted little time to subduing fortifications in the main text, and the select list suffers from being

---

1   Carl Von Clausewitz, *On War*, Book 7, Chapter 17, 'Attack on Fortresses', available at <https://clausewitzstudies.org/readings/OnWar1873/BK7ch17.html>, Accessed 14 November 2024.
2   Ilya Berkovich, *Motivation in War: The Experience of Common Soldiers in Old-Regime Europe* (Cambridge: Cambridge University Press, 2017), p.17; M.S. Anderson, *War and Society in Europe of the Old Regime, 1618–1789* (Avon: Leicester University Press, 1988), p.197.
3   Clausewitz, *On War*, Book 7, Chapter 17, 'Attack on Fortresses', available at <https://clausewitzstudies.org/readings/OnWar1873/BK7ch17.html>, Accessed 14 November 2024.
4   Michael Glover, *Wellington as Military Commander* (London: B.T. Batsford, 1968), pp.162–177.

Madeira: *The Maratha and Jat Campaigns 1803-1806*, and *Every Hazard and Fatigue: The Siege of Pensacola 1781*. Historical military operations against fortified places have been his focus for several years.

Dr Silvia Gregorio-Sainz is a lecturer in the English Department at the University of Oviedo. She holds a PhD in English Studies, as well as an MA in Socio-Cultural Analysis and History and an MA in TEFL. Her research focuses on Anglo-Spanish relations in the first half of the nineteenth century, particularly during the Peninsular War and its aftermath. Recent publications include the book *La Guerra de la Independencia en Cantabria (1808-1813): una visión británica* (Editorial Universidad de Cantabria, 2025). She is also a member of the ongoing international project EURICAR'20, which examines European cultural reactions to the Spanish First Carlist War (https://www.uniovideo.es/euricar20/).

Mark S. Thompson is an independent historian specialising in the Peninsular War 1808-1814. His first book, *The Fatal Hill* (2002), covered the 1811 Albuera campaign. Other books include: *Wellington's Engineers* (2015), which was based on his PhD; *Wellington's Favourite Engineer* (2020) about the services of Field Marshal John Burgoyne RE; and *Wellington and the Lines of Torres Vedras* (2021). Mark has published ten other books on the period. He has presented at many conferences in the UK and abroad, including the Wellington Congress, and is a member of the British Commission for Military History, the Friends of the British Cemetery at Elvas, and the Friends of the Lines of Torres Vedras.

Dr Zack White is an award-winning historian, broadcaster and Leverhulme Early Career Research Fellow at the University of Portsmouth. The host of the Napoleonic Wars Podcast, and founder of the Napoleonic & Revolutionary War Graves Charity, he has appeared on the BBC, Sky and Channel 4 sharing his knowledge and passion for the period. He previously published *The Sword & Spirit: Proceedings of the First War and Peace in the Age of Napoleon Conference* with Helion in 2021.

Contents

Contributor Biographies — iv

Introduction: The Neglected Yet Unavoidable Evil of Siege Warfare in the Age of Napoleon — vi
Zack White

1. Siege Warfare During the Russo-Ottoman War of 1806-1812: A Case Study of the Assault on Brăila — 11
Alexander Mikaberidze

2. The Siege of Izmail, December 1790: Russian Military Culture in the Early Revolutionary Period — 30
Eugene Miakinkov

3. Forging a Chain of Security. The Story of the Wellington Barrier, 1815–1830 — 52
Beatrice de Graaf

4. The Logistics of a Successful Siege: The Planning for the Allied Siege of Ciudad Rodrigo, January 1812 — 72
Mark S. Thompson

5. The Siege Lords: British Siege Warfare in India, 1792–1805 — 93
Joshua Provan

6. The Bombardment of Antwerp, February 1814 — 113
Andrew Bamford

7. Revisiting British Accounts: The 'Other' Deadly Siege on the Northern Coast of Spain (1813) — 139
Silvia Gregorio Sainz

Contributor Biographies

Dr Andrew Bamford obtained his PhD at the University of Leeds for a study of the British Army's regimental system, which was subsequently adapted for publication as *Sickness, Suffering, and the Sword* (University of Oklahoma Press, 2013). He has subsequently written extensively on the British Army in the Napoleonic era as well as the wars of the 1740s, making a particular study of the Netherlands campaign of 1813–1814, and was founding editor of Helion's *From Reason to Revolution 1721-1815* book series. He is also the co-host, with Dr Alexander S. Burns, of the *Prime and Load* Podcast, which deals with warfare in the eighteenth century.

Professor Beatrice de Graaf is a distinguished professor and has held the chair of the History of International Relations at Utrecht University since 2014. She has published widely on the history of security, crisis, conflict and terrorism. Her CUP monograph *Fighting Terror After Napoleon* (2020) was awarded the Arenberg Prize for the Best Book in European History (2022). Amongst others, she was a fellow at Catherine's College Cambridge (2016, 2024), is a fellow at CASSIS/Bonn University, and a member of the Royal Netherlands Academy for Arts and Sciences. She is the Core Editor of the *Journal for Modern European History and of Terrorism and Political Violence*.

Dr Eugene Miakinkov is a Senior Lecturer in History at Swansea University. He specialises in the history of Imperial Russia, especially the eighteenth century. He is the author of *War and Enlightenment in Russia: Military Culture in the Age of Catherine II* (University of Toronto Press, 2020). He is currently working on a project about the role of militarism in Russian history.

Alexander Mikaberidze is Professor of History and Ruth Herrin Noel Endowed Chair at Louisiana State University-Shreveport. He has written and edited over two dozen books, including *Kutuzov: A Life in War and Peace,* and the award-winning *The Napoleonic Wars: A Global History* (2020), which has been translated into five languages. He has served as an editor of the multi-volume *Cambridge History of the Napoleonic Wars* (2023).

Josh Provan is an independent historian and writer whose research focuses on global culture and conflict. He is the author of three books on eighteenth- and nineteenth-century British colonial history, as well as various papers, articles and chapters covering a much wider period and subject matter. His published work to date includes *Wild East: The British in Japan 1853-1868, Bullocks Grain and Good*

largely western-European-centric in its selection.⁵ More recent thought-provoking and important volumes, such as Christy Pichichero's *The Military Enlightenment* and Roger Chickering and Stig Förster's *War in an Age of Revolution* have considered sieges to be beyond the reasonable scope of their analysis.⁶

Of course, scholars have by no means ignored sieges entirely, but consideration of them has undoubtedly been sporadic, and often primarily undertaken from an operational perspective or in the form of campaign histories. The most comprehensive coverage in this style has been given to the sieges of the Peninsular War campaigns. Studies such as Bruce Collins's work on the siege of San Sebastian and Tom Saunders's book on Ciudad Rodrigo have added to our knowledge of individual case studies. Through this approach, it is possible to tangentially gain an appreciation of the logistical challenges of attempting a siege, as well as the toll on the soldiers. However, a more holistic consideration of how sieges worked and why they mattered during this period has been missing.⁷ Frederick Myatt's volume *British Sieges of the Peninsular War*, while offering some useful thoughts on the plundering that took place in the wake of these events, has obvious limitations in its scope, as the title implies. The British were by no means the only ones conducting siege operations in the theatre, and in any case, these events need to be set in a much wider strategic, political and social context.

The one place where significant progress has been achieved is in relation to soldier and civilian experiences, both during and after sieges, though again the focus has predominantly been on British operations. Poor discipline amongst troops after fortresses were stormed has particularly been a focus of study here, with Ed Coss setting the tone for much of this discussion by highlighting how the sackings were exacerbated by soldiers' expectations that they could plunder as a reward for hard fighting, and peer pressure from their colleagues.⁸ The editor of this volume has also studied the legal implications of the sackings, finding that British troops universally avoided prosecution for sacking the fortress cities of Ciudad Rodrigo and Badajoz in 1812.⁹ Gareth Glover's *Marching, Fighting, Dying* is an important contribution to appreciation of what it was like to experience a siege, drawing extensively on soldiers' accounts to highlight the experience of living through and fighting in a siege.¹⁰ This approach has been taken further by Gavin Daly, whose 2022 publication *Storm and Sack* is the most important single title dedicated to siege warfare in

---

5   Gunther Rothenberg, *The Art of Warfare in the Age of Napoleon* (Bloomington: Indiana University Press, 1978), pp.254–255.
6   C. Pichichero, *The Military Enlightenment: War and Culture in the French Empire from Louis XIV to Napoleon* (Ithaca: Cornell University Press, 2021); R. Chickering & S. Forster (eds), *War in an Age of Revolution, 1775–1815* (Cambridge: Cambridge University Press, 2010).
7   Bruce Collins, *Wellington and the Siege of San Sebastian* (Barnsley, Pen and Sword, 2017); Tim Saunders, *The Sieges of Ciudad Rodrigo 1810 and 1812* (Barnsley: Pen and Sword, 2018).
8   Edward Coss, *All for the King's Shilling: The British Soldier under Wellington, 1808–1814* (Norman: Oklahoma University Press, 2010).
9   Zack White, 'Pragmatism and Discretion: Discipline in the British Army, 1808–1818' (Unpublished PhD Thesis, University of Southampton, 2022), pp.147–153.
10   Gareth Glover, *Marching, Fighting, Dying: Experiences of Soldiers in the Peninsular War* (Barnsely: Pen & Sword, 2021), pp.225–254.

the Napoleonic era. Daly's work sets British actions in the aftermath of sieges within a much wider context of eighteenth-century discussions on the rules of warfare and acceptable behaviour. In doing so, he highlights the incompatibility of eighteenth-century enlightened and mathematical approaches to war with the emotional turmoil of soldiers' experiences in sieges.[11]

It is therefore clear that the vast majority of work looking at sieges in the Napoleonic era is dominated by an Anglo- and Peninsular-War-centric focus. Yet over the last decade and a half, the study of the Napoleonic Wars more broadly has made significant progress in many areas, as social, legal, gender and cultural history have become an ever-more integral part of key discussions and debates.[12] Alex Mikaberidze's work has set the conflict within a global context, providing a timely reminder of the need to take a more holistic view of military history, both in terms of geographic scope, and its implications.[13] It is important to respond to these developments by bringing the same mentality to bear on the discussion of siege warfare in the Napoleonic Wars. Sieges, just like pitched battles, did not exist in isolation. Fortresses were intrinsically linked to wider discussions of diplomacy, and any decisions taken on which towns should be fortified were deeply political, as well as practical and strategic, considerations. From sourcing the materiel required to build or breach them, to shaping the livelihoods of those who lived in their shadows, fortresses caused lasting changes to the landscape by impacting both the physical geography and local economy. It would also be a mistake to assume that thinking on how to approach sieges stood still in an age dominated by innovations in the way war was waged. This volume seeks to develop our recognition of these and other factors in a wider-ranging exploration of case studies which highlight innovations in tactics, the political dimensions to building and storming fortresses, and widen the conversation in terms of geographic scope when we discuss sieges in the Napoleonic era.

Underpinning the need to move away from an exclusively Peninsular War-focused narrative, the first two chapters consider the Russian experience of sieges during the period. Alex Mikaberidze's analysis of sieges conducted during the Russo-Ottoman War of 1806–1812, in Chapter 1, highlights the fact that sieges continued to have an important role to play in the conduct of campaigns. Using the case study of the Russian assault on Brăila in 1809, he highlights the wider impact that an unsuccessful storming could, but ultimately did not, have in terms of honing strategy.

Eugene Miakinkov continues the volume's examination of siege operations in Eastern Europe with a detailed account of the siege of Izmail in 1790, offering a

---

11 Gavin Daly, *Storm and Sack: British Sieges, Violence and the Laws of War in the Napoleonic Era, 1799–1815* (Cambridge: Cambridge University Press, 2022), p.270.
12 See for example Beatrice de Graaf, *Fighting Terror after Napoleon: How Europe became Secure after 1815* (Cambridge: Cambridge University Press, 2020); Thomas Malcolmson, *Order and Disorder in the British Navy: Control, Resistance, Flogging and Hanging* (Martlesham: Boydell Press, 2016); Karen Hagemann, Gisela Maletta and Jane Rendall (eds), *Gender, War and Politics: Transatlantic Perspectives, 1775–1830* (Basingstoke: Palgrave Macmillan, 2010); Luke Reynolds, *Who Owned Waterloo?: Battle, Memory and Myth in British History* (Oxford: Oxford University Press, 2022).
13 Alex Mikaberidze, *The Napoleonic Wars: A Global History* (Oxford: Oxford University Press, 2020).

different interpretation of the degree to which Russian commanders learnt from bloody siege assaults. By examining Russian military culture during the period, he also offers insights on the extent to which different nationalities and, perhaps, national characters, may have utilised different approaches to siege warfare, a concept which Josh Provan's chapter also alludes to.

There is much more to siege warfare than simply reducing fortifications, launching assaults and trying to secure surrenders, and Beatrice de Graaf's chapter on the 'Wellington Line' of fortifications provides a useful reminder of the impact that these buildings could have in times of peace. Considering everything from environmental degradation to corruption, her work highlights how these places impacted those living in their shadow, and offers a refreshing take on this volume's theme that sieges do not exist in isolation.

Highlighting innovative approaches and experimentation in warfare is equally important, yet is often overlooked in the context of sieges. Whilst bombardment was not a revolutionary new aspect of conflict in this period, generals were nonetheless unsure on how best to utilise it when considering siege operations. In Chapter 4 Andrew Bamford highlights that uncertainty, drawing connections between the results of the bombardment of Copenhagen and the overly ambitious planning for the bombardment of Antwerp.

A global perspective to siege operations is provided in Chapter 5 by Josh Provan's chapter as he considers the British approach to sieges in India. In doing so, he highlights how integral the swift reduction of fortifications was to Britain's way of waging war on the subcontinent. In a careful consideration of a wide range of case studies, it becomes clear that overwhelming aggression was often resorted to in order to maintain an impression that there was nothing which could withstand the combined might of King George III's men and the East India Company's sepoys.

Although this introduction has emphasised the need to move away from a Peninsular War-centric discussion of siege warfare, this is not to imply that we cannot learn valuable lessons from its history by approaching the events of the conflict with fresh questions. Mark S. Thompson's exploration of the logistics of sieges in Chapter 6 is a perfect example of this, offering a detailed understanding of the planning these operations required and how easily plans could be derailed.

The importance of dispensing with an Anglo-centric narrative of the Peninsular War has long been noted by historians.[14] This volume's final chapter by Silvia Gregorio Sainz uses the siege of Castro Urdiales to re-emphasise this point, drawing our attention to the brutality that the French showed Spanish civilians following their capture of the fortress town in May 1813. This is a little-known siege that has been largely ignored in wider histories of the period, in part due to Anglocentrism. Her consideration of the town's wider role in the war, and British interest in the region, highlights how sieges can provide engaging and pertinent 'micro-studies' through which we can appreciate the complex, yet often-overlooked, day-to-day business of war.

---

14   Charles Esdaile, *The Peninsular War: A New History* (London: Penguin, 2003).

Taken together, these chapters offer one of the broadest surveys of siege warfare in the Napoleonic age that has ever been attempted. No attempt has been made to be comprehensive in the coverage, not least because to do so would require a much longer volume than was feasible. Nonetheless, this compendium offers a starting point from which enthusiasts and historians of the period can begin to think more broadly about the impact of sieges. More than anything else it reaffirms the judgement of Clausewitz that sieges were both evil and unavoidable – an indispensable part of conflict in the Napoleonic era, which widens our appreciation of its complexities, and its horrors.

I cannot close this editorial without taking the time to thank a number of people who have made this volume happen. Andrew Bamford was the individual who first invested in the idea underpinning this volume, and as part of his wider, revolutionary ethos as series editor at Helion, he was willing to back the concept behind this work when others would not have given it the time of day. Rob Griffith has since taken up the torch, and has been a patient, insightful and sensitive editor, who has made the challenges of compiling an edited collection immeasurably more bearable every time I have worked with him. This volume was delayed by a year due to challenging personal circumstances, and the fact that it ever came to be completed was thanks, in part, to the kindness of countless people. There are too many for me to name everybody, but I would particularly like to thank the loyal fans of *The Napoleonic Wars Podcast* for their encouragement. I am especially fortunate to be surrounded by a group of dear and supportive friends and family to whom I owe an unrepayable debt. My love and thanks are especially owed to Rosie White, Rory Muir, Ed Coss, Beatrice de Graaf, Graeme Callister, Peter Molloy, Luke Reynolds, and Liam Telfer. Above all, I would like to thank my long-suffering authors, who have shown phenomenal patience in the face of significant delays. Their commitment to the vision underpinning this volume and their willingness to share their eye-opening research have been a source of great inspiration as I have worked on this book. Without them, the conversation and debate could not progress.

Zack White
University of Portsmouth
January 2025

1

Siege Warfare During the Russo-Ottoman War of 1806-1812: A Case Study of the Assault on Brăila

Alexander Mikaberidze

In popular imagination, the Napoleonic Wars are associated with the decisive battles. For many, the names of Austerlitz, Salamanca, Leipzig, and Waterloo have become synonymous with the turbulent first two decades of the nineteenth century. Few think of siege warfare playing an important role during this period. Yet, the sieges continued to play an important role; young Napoleon Bonaparte himself had seen his career launched by his performance at the Siege of Toulon in the autumn of 1793. More crucially, the Napoleonic period is replete with examples of large-scale sieges that shaped the course of the wars, especially in the Iberian Peninsula.[1] Although proportionally less numerous than before, these sieges were ferociously contested and produced some of the worst carnages of the entire era. This is especially true of the Russo-Ottoman Wars in the Danubian Principalities and the Caucasus.

The eighteenth century saw the escalation of the long-simmering hostilities between the rising power of the Russian Empire and the Ottoman Empire, which struggled to maintain its traditional hegemony in southeastern Europe. The two powers fought in 1710–1713, 1735–1739, 1768–1774, and 1787–1791. The later conflicts witnessed the Russian army consistently prevailing over its Ottoman opponents in decisive battles but then bogging down in reducing the defensive line of Ottoman fortresses, first in Ukraine and then in the Danubian Principalities. The Russo-Ottoman war of 1787–1791 is particularly interesting in this regard. Of the 10 pivotal land engagements of this war, only three were battles (Focșani and Rymnik in 1789, Măcin in 1791), the remaining seven involved sieges and combats resulting from attempts to relieve them:

---

1 See Gavin Daly, *Storm and Sack: British Sieges, Violence and the Laws of War in the Napoleonic Era, 1799–1815* (Cambridge: Cambridge University Press, 2022).

Table 1:

Place	Date	Event
Kinburn	October 1787	A small fortress located opposite the much larger Ottoman fortress of Ozi (Ochakov), Kinburn covered approaches to the fleet base at Kherson. In October 1787, the Ottomans landed at Kinburn to deprive the Russians of a base for the impending siege of Ochakov.
Ochakov	May–December 1788	Siege of the Ottoman fortress by the Russian army led by Prince Gregory Potemkin and General Alexander Suvorov. The fortress was stormed in December 1788.
Chocim/ Khotin	May–September 1788	Siege of the Ottoman fortress by Russo-Austrian fortress; the Ottoman garrison fell in September 1788.
Belgrade	September–October 1789	Siege of the Ottoman-held fortress by a Habsburg Austrian army led by *Feldmarschall* Ernst Gideon von Laudon. The fortress fell in October but the Ottoman garrison was allowed a free passage.
Cetingrad	June–July 1790	Engagement between the Croatian Corps of the Habsburg army, led by *Feldzeugmeister* Joseph Nikolaus Baron de Vins, and the Ottoman army, led by Dizdar-Agha Ali-Bey Beširević, during the siege of the Cetin Castle in central Croatia.
Izmail	November–December 1790	Siege and storming of the strategically important Ottoman fortress on the lower Danube by the Russian army led by Alexander Suvorov. The fortress was stormed in December, with much of its garrison and population massacred. The song *Grom pobedy, razdavaysya!* (Let the thunder of victory sound!) written to celebrate the Russian victory at Izmail served as an unofficial Russian anthem until 1833.
Anapa	July 1791	Siege and storming of the Ottoman fortress of Anapa on the northeastern shores of the Black Sea by the Russian army led by General Ivan Gudovich.

The Russian assaults of Ochakov and Izmail have been widely publicised (and embellished), but less well-known is the Russian struggles to reduce other Ottoman fortresses, often with calamitous results, in subsequent years. This paper examines the siege warfare during the largely overlooked Russo-Ottoman War of 1806–1812, with a particular focus on the disastrous assault on the fortress of Brăila that revealed the challenges of conducting siege warfare in the Age of Napoleon.

At the start of the nineteenth century, the condition of the Ottoman Empire was so dire that some were predicting its dissolution. In the Caucasus, Russia had successfully supplanted the Ottoman influence in the Georgian principalities, first annexing the eastern Georgian kingdom of Kartli-Kakheti in 1801 and then extending its dominion to the Kingdom of Imereti and the Principalities of Megrelia and Guria, which the Ottomans traditionally claimed as spheres of interest. Elsewhere, the sultan's authority was routinely disregarded: Pasvanoglu Osman Pasha, the governor of Vidin, was in open revolt, as was almost all of Serbia; the *hospodars* of Moldavia and Wallachia showed clear affinities for self-rule while Ali Pasha of Janina was obedient only when it suited his convenience. In the empire's eastern provinces, Djezzar Pasha enjoyed virtual independence in Syria, and Mehmed Ali was zealously consolidating his power in Egypt, now that the French had shattered the power of the Mamluks; meanwhile, the Wahhabi fundamentalists extended their influence across much of Arabia. The Ottoman realm was truly, as historian Virginia Aksan aptly put it, 'an empire besieged.'[2]

The Russo-Ottoman relations, strained as they were, worsened in 1804–1805 when Emperor Napoleon approached Sultan Selim III with a proposal of a possible alliance. The possibility of French presence in the Balkans and/or control of the strategic straits of the Dardanelles and Bosporus had concerned Alexander, who sympathised with the plight of the Slavic peoples under Ottoman domination and extended support to the Serbian insurgents. Caught between a rock (France) and a hard place (Russia), the Ottomans initially sought to remain neutral and avoid entanglement in European squabbles but, in the wake of Napoleon's triumph at Austerlitz in December 1805, cast their lot with the French side. Sultan Selim III acknowledged Napoleon's imperial title, lent a willing ear to the talk of an alliance with France to confront Russia, and, most crucially, replaced the current pro-Russian *hospodars* of Moldavia and Wallachia with more pliant candidates. These initiatives threatened to overturn the existing state of affairs between Russia and the Ottoman Empire. Czar Alexander and his advisers were quick to point out that the Ottoman decision to remove the *hospodars* had violated the articles of the Treaty of Jassy of 1792, which required joint agreement when it came to dismissing or appointing them. As negotiations over the future of Moldavia and Wallachia reached a breaking point, both sides began concentrating their troops on their borders. In October 1806, just as he began preparing for the War of the Fourth Coalition, Alexander ordered the invasion and occupation of the Danubian Principalities. In just three months, General Ivan Michelson, with some 40,000 men, had overrun these territories and driven the Turkish forces back to the Danube where the mighty river, buttressed by a line of formidable Ottoman fortresses, contained the Russian momentum.

Over the next three years, the war in the Danubian valley followed a familiar pattern of the Russian forces successfully confronting the Ottoman counterattacks in the field – Turbat, Giurgiu and Obilesti in 1807 alone – but unable to launch a trans-Danubian offensive due to logistical challenges and persistent threats from the garrisons of Ottoman fortresses on the Danube River. In 1807–1808, the Russian forces tried to seize these fortresses, but the results were mixed. The poorly-defended fortresses of Khotin, Bender and Akkerman fell in 1807, but the great fortress of Izmail withstood the Russian siege in March–July 1807. Furthermore, the Russian campaign on the Danube was obstructed by the ongoing conflict between Russia and Napoleonic France in Europe, which required diverting much-needed forces from the Danubian theatre. Indeed, Russian victories over the Ottomans were seemingly offset by the defeats in Poland until the Peace Treaty of Tilsit, negotiated in July 1807, gave the Russians a chance to consolidate their positions in the Ottoman realm.

* * *

---

2 See Virginia Aksan, *The Ottomans, 1700–1923: An Empire Besieged* (New York: Routledge, 2021)

14 AN UNAVOIDABLE EVIL

The Danubian provinces.

Unaware of these momentous political changes, General Ivan Meyendorff, who replaced Michelson as the commander of the Russian forces in the Danubian Principalities, signed an armistice with the Ottomans at Slobozia in late 1807. The agreement, negotiated without Czar Alexander's sanction, called for Russian withdrawal from Moldavia and Wallachia while the Turks agreed to remain south of the Danube.[3] Enraged by the general's decision, the czar dismissed him and sent Prince Alexander Prozorovskii as the new commander-in-chief of the Army of Moldavia.

The new commander's immediate task was to inform the Ottomans that the armistice was void because his predecessor had had no authority to accept it. The Ottomans objected, and the two sides spent the rest of the year debating the legality of the agreement; neither wanted to rush into a resumption of hostilities before being fully prepared. Two years of military operations and the presence of tens of thousands of troops had taken a toll on the Danubian regions, which suffered from widespread looting; regular outbreaks of scurvy, plague, and other diseases; and the exodus of thousands of peasants, with the resultant drop in agricultural production.

The Army of Moldavia, which listed some 80,000 men on its rosters, suffered from a lack of supplies and adequate logistical support to conduct decisive military operations.[4] The army consisted of several corps. General Mikhail Miloradovich led one on the left flank, near Bucharest, while the advance guard, under Cossack Ataman Matvei Platov, stood near Rymnik (present-day Râmnicu Sărat, Romania). On the right wing was Count Alexander Langeron's corps, supported by a special detachment that kept an eye on the Turks at Galati.[5] The army's central force, the so-called Main Corps, was placed under General Mikhail Golenischev-Kutuzov's command and represented a sizable force of 19 infantry battalions, 30 squadrons of cavalry, two heavy artillery companies, and one horse artillery company, supported by several hundred Cossacks and companies of pioneers and pontoniers. Moreover, during the summer of 1808, Kutuzov's corps was reinforced with thousands more troops from Peter von Essen III's 8th Division and General Zakhar Olsufyev III's 22nd Division.

---

3   For a wide context of the war, see A. Petrov, *Voina Rossii s Turtsiej 1806–1812 gg.* (St Petersburg: V. S. Balashev, 1887), vol.II, pp.1–44; Alexander Mikhailosvkii-Danilevskii, *Polnoe sobranie sochinenii. Tom III: Opisanie Tureckoi voiny s 1806 do 1812 goda* (St Petersburg: Tip. Shtaba Otd. Korpusal Vnutren. Strazhi., 1849), pp.1–82. Unfortunately, there is still no modern English-language study of the Russo-Ottoman Wars of 1787–1791 and 1806–1812. For an earlier conflict, see Brian Davies, *The Russo-Turkish War, 1768–1774. Catherine II and the Ottoman Empire* (London: Bloomsbury, 2016). For a wider context, see Brian Davies, *Empire and Military Revolution in Eastern Europe: Russia's Turkish Wars in the Eighteenth Century* (London: Continuum, 2011). For the Ottoman side, see Virginia H. Aksan, *Ottoman Wars, 1700–1870: An Empire Besieged* (New York: Pearson, 2007); Stanford J. Shaw, *History of the Ottoman Empire and Modern Turkey* (Cambridge: Cambridge University Press, 1976), vol.I, pp.217–276.
4   For details, see Disposition of the Army of Moldavia, in *M. I. Kutuzov: Sbornik dokumentov*, edited by Liubomir Beskrovnyi (Moscow: Voeniszdat, 1950), vol.III, pp.20–29.
5   Disposition of the Army of Moldavia, in *M. I. Kutuzov: Sbornik dokumentov*, vol.III, pp.20–29.

As always, logistics was the most pressing issue. During the earlier campaigns, Russian supplies had to be carried from the distant bases in the provinces, and transporting vast amounts of ammunition and provisions over such long distances was costly and time-consuming. To expedite the process, the Russian headquarters first undertook an inventory of supplies in the army and the neighbouring Russian provinces. Kutuzov established a central depot to which the Moldavian and Wallachian authorities were now expected to deliver supplies; smaller depots were set up in various towns and villages, and arrangements were made to purchase food directly from the local population. Throughout the spring and summer of 1808, the Russian military authorities reorganised the field hospital system to accommodate sick soldiers, took measures to combat scurvy (which spread due to the lack of fresh supplies), and organised new medical facilities, including an officers' hospital in Jassy, a mobile infirmary, and a medical depot for soldiers. They made sure soldiers received new knapsacks, coats, and other equipment that had been recently modified.[6] Equally important were Russian efforts to train the new recruits for the impending campaign. The 8th and 22nd Divisions, which arrived in late summer, had thousands of inexperienced recruits who had to be properly prepared before the campaign started. Special training camps were set up, where regiments could be regrouped, reinforced, and trained.[7]

The Russian preparations were made possible by the mayhem unfolding in the Ottoman Empire. Sultan Selim III was beset by too many problems to think about resuming hostilities. In May 1807, the Janissaries erupted in a rebellion that forced the sultan to scale back his designs and disband the recently formed Nizam-i Cedid regiments that had been trained and equipped in the European manner. But even such drastic concession could not save the sultan – in late May 1807 the Janissaries dethroned Selim and replaced him with young and inexperienced Mustafa IV, thereby ushering in a political crisis that hampered Ottoman military capabilities and forced the empire's armies, some of whom were commanded by provincial notables engaged in power struggles, to seek a defensive posture on the Danube. The Ottoman political crisis raged over the next 12 months as Bayraktar Mustafa Pasha, the powerful governor of Ruse, marched on Istanbul in July 1808 to reinstate Selim. He arrived too late to save the former sultan, who was murdered by the rebels, but in time to depose Mustafa and install Mahmud II on the Ottoman throne in late July 1808. Like his predecessor, Mahmud II was politically impotent and depended for his survival on the outcome of the crisis that raged with ferocious intensity. Bayraktar Mustafa Pasha revived some of Selim III's modernising reforms, engendering a strong opposition from the religious clerics, Janissaries, and conservative elements of Ottoman society. A new revolt flared up in November 1808 in Istanbul, with the rebels killing the influential pasha and many of his supporters.

---

6   Kutuzov to Minister of War A. Arakcheyev, April 25/May 7, 1808, in *M. I. Kutuzov: Sbornik dokumentov*, vol.III, pp.5–6.
7   Kutuzov's Orders to the Army, May 10/22 and May 26/June 7, 1808, Prozorovskii to Kutuzov, July 7/19, 1809; Kutuzov to Essen III and Olsufyev III, 9/21 July 1809, in *M. I. Kutuzov: Sbornik dokumentov*, vol.III, pp.14, 30–31, 40–42.

In light of this political turmoil, the Ottomans were unsurprisingly very keen on preserving the Armistice of Slobodzia with Russia. Czar Alexander, however, had little desire for peace at a moment when his Ottoman enemy was in such disarray, and while Napoleon was preoccupied with the Spanish affairs. 'The Ottoman Empire is dead,' the Russian foreign minister told the French ambassador, 'so why should not Russia keep the spoils of war?'[8] Alexander increasingly thought in expansionist terms, seeking to compensate for what he had lost to Napoleon's aggrandizement in Central Europe by taking territory in the Balkans. In October 1808, the French and Russian emperors met once more at Erfurt with the goal of consolidating their alliance. In exchange for his support for France's war against Austria, Alexander extracted Napoleon's acceptance of the Russian presence in the Danubian Principalities and a promise not to interfere in Russian relations with the sultan.[9]

Upon returning from Erfurt, Alexander instructed Prozorovskii to present an ultimatum to the Ottomans calling for them to surrender the Danubian principalities or to face the immediate resumption of hostilities.[10] Sultan Mahmud, predictably, refused to consider the terms, and the Russians prepared for the new offensive. With Napoleon preoccupied with wars in Spain and Austria, the moment seemed right to consolidate Russian interests in southeastern Europe. 'A swift crossing of the Danube is the only sure way to force the sultan to accept peace and surrender of Bessarabia, Moldavia, and Wallachia,' Alexander told Prozorovskii.[11]

The Russian commander-in-chief knew that the Ottomans had some 40,000 men under Grand Vizier Yusuf Pasha deployed between Adrianople and the Danube; continued disturbances in Istanbul hampered Ottoman logistics and mobilisation efforts, with many local governors reluctant to provide additional troops. This meant, or so the Russian commander hoped, that the field army would be immobilised for some time. However, Russian intelligence also reported that the Turks had about 50,000 men scattered in fortresses along the Danube, with about 12,000 men at Brăila, 10,000 at Izmail, and thousands more defending entrenchments at Giurgiu, Ruse, Silistra, Turnu, Vidin, and other places. Despite the czar's call for the Russian army to immediately cross the Danube River, Prozorovskii felt uneasy about leaving such strong enemy garrisons in his rear. He declined, therefore, to conduct a rapid trans-Danubian campaign to seek out and destroy the Ottoman field army and instead insisted on the systematic reduction of the enemy fortresses, starting with those of Giurgiu and Brăila. Once the Danube's northern bank was

---

8   Savary to Napoleon, November 4, 1807, in *Sbornik Imperatorskago russkago istoricheskago obschestva* [cited as *SIRIO*] 83(1892), p.180.
9   See Articles 8, 9, and 11 of the Erfurt Convention, <http://www.napoleon-series.org/research/government/diplomatic/c_erfurt.html.>
10 For details see *Vneshnaya politika Rossii XIX I nachala XX veka: dokumenti Rossiiskogo Ministerstva Inostrannikh del* (Moscow: Gos. izd. polit. lit., 1960), vol.IV, pp.365–369, 439–440, 456–458; Count Rumyantsev to Prince Prozorovskii, August 7 [19], 1808, and October 3 [15], 1808, Weimar, cited in Mikhailovskii-Danilevskii, *Opisanie Turetskoi voiny s 1806 do 1812 goda*, pp.88–89.
11 Cited in Mikhailovskii-Danilevskii, *Opisanie Turetskoi voiny s 1806 do 1812 goda*, p.104.

secured, the Russian army[12] would cross the river and target the Ottomans in the Bulgarian uplands.[13]

When drafting the new operational plan, Prozorovskii had decided to first attack with his right wing, ordering Miloradovich to capture Giurgiu. In April, as the Russians approached this fortress, its Ottoman garrison proved to be well prepared. Miloradovich soon realised (as Alexander Suvorov had at Izmail in 1790) that he had miscalculated the depth of the moat surrounding the fortress and that the Ottomans had improved their defences. Precious time was lost as the Russian generals scrambled to regroup their attack columns. Under heavy enemy fire, casualties mounted so rapidly that the Russian commander had no choice but to beat a retreat, losing some 700 men in the process.[14]

Dismayed by the setback at Giurgiu, Prozorovskii now turned to the fortress of Brăila, an imposing castle overlooking the Danube on one of the river's last major bends before it empties into the Black Sea. Kutuzov was tasked with leading the Main Corps to the fortress.[15] The Main Corps, some 35,000 strong, left Focșani on April 12 and marched toward the Danube, some 60 miles away.[16] The weather seemed to conspire against the Russians; cold, torrential rains hindered their advance.[17] Some units lost their way, forcing Kutuzov to halt the march for several days so he could rally his forces on the Buzau River, about 20 miles from Brăila. Leaving a small rear guard and sending forth scouts to reconnoitre the area, Kutuzov then led the Main Corps toward Brăila in late April, brushing aside the Turkish outposts and initiating the blockade of the fortress. The Russians deployed in a wide arch around the fortress, with General Sergey Kamenskii's division forming the right wing, Essen III's units on the left, and Yevgeny Markov's troops in the centre.[18]

* * *

---

12 The Army of Moldavia had spent the winter of 1809 in cantonments scattered across Wallachia. Miloradovich's right wing corps was near Bucharest, while General Langeron's corps held positions in front of Izmail on the left wing. Kutuzov's Main Corps stayed in between them, guarding a vast area from Jassy to Focșani. The main corps comprised five divisions: Essen III's 8th Division was at Byrlad (Bârlad), Olsufyev III's 22nd Division was at Bacâu, Rtischev's 16th Division was at Vaslui, Kamenskii's 12th Division was at Focșani, and Markov's 15th Division was at Roman. See Disposition of the Army of Moldavia, in *M. I. Kutuzov: Sbornik dokumentov*, vol.III, pp.76–80. The army entered the winter cantonments in late November. See Kutuzov to Prozorovskii, November 7/19, 1808, in *M. I. Kutuzov: Sbornik dokumentov*, vol.III, pp.82–84.

13 One eminent Russian military historian criticized Prozorovskii for approaching the planning of the new campaign with 'the tactics of the 1769 campaign' still on his mind. A. Petrov, *Vlianie Turetskikh voin s polovini proshlogo stoletia na razvitie Russkago voennago iskusstva* (St. Petersburg, 1894), p.227.

14 Mikhailovskii-Danilevskii, *Opisanie Turetskoi voiny s 1806 do 1812 goda*, pp.106–107; Petrov, *Voina Rossii s Turtsiej 1806–1812 gg.*, vol.II, pp.206–215.

15 Prozorovskii to Kutuzov, March 28/April 9, 1809, in *M. I. Kutuzov: Sbornik dokumentov*, vol. III, pp.118–120.

16 See Kutuzov's orders of April 10–12 and the marching disposition of April 11 in in *M. I. Kutuzov: Sbornik dokumentov*, vol.III, pp.126–131.

17 Afanasii Krasovskii, 'Iz zapisok general-adjutanta Krasovskago,' in *Russkii Vestnik* 8 (1880), p.502.

18 For details, see the Journal of Military Operations of the Main Corps in *M. I. Kutuzov: Sbornik dokumentov*, vol.III, pp.141–143, 150–151. Also see Kutuzov's Marching Disposition of 6/18, April 1809, in *M. I. Kutuzov: Sbornik dokumentov*, vol.III, pp.144–146.

The fortress of Brăila was built around a former Orthodox monastery. The pentagon-shaped citadel had been overhauled in recent years, its stone walls repaired, and its towers reinforced. Jutting from the walls were 35-foot-tall diamond-shaped bastions that allowed the Ottomans supporting fields of fire. The bastions were further protected by an extensive moat. Nonetheless, most of the garrison's 12,000 men were stationed outside the citadel, protecting the retrenchment – a massive secondary line of defence consisting of a deep and wide ditch, with a steep counterscarp that the attacking units would have to descend before reaching the earthen ramparts with five bastions, which had been constructed within the range of the citadel's cannon. To capture the castle, the Russians would have to overcome these earthworks first.

Kutuzov commenced the siege works on 19–20 April. Over the next few days, the Russian engineers, led by Engineer Major General Ivan Harting, constructed parallels that approached within half a mile of the retrenchment, as well as four large breaching batteries to target the Ottoman positions. The garrison made daily sorties but failed to interrupt the Russian preparations; instead, the Turks set the outer suburbs on fire to clear the view and to deny the enemy any cover.[19] On April 23, Russian heavy cannon arrived, along with a flotilla of 19 gunboats from Galati. The artillery barrage silenced Ottoman artillery and caused significant damage in the Brăila suburbs.[20] By the end of April, Kutuzov's disciplined reduction of the fortress seemed to have been working: Russian gunboats controlled the river, while the Main Corps effectively isolated the fortress on land.[21]

In early May, Field Marshal Prozorovskii arrived to assume the overall command of the corps. Now that Kutuzov had set up the blockade, the main question was whether to besiege the fortress or to expedite the process by a direct assault. Emboldened by the Russian artillery bombardment and the apparent feebleness of the Ottoman defences, Prozorovskii thought the fortress would be an easy target, something to be taken, as he put it, 'on the fly' (*mimokhodom*). He thus ordered the assault.[22]

Yet senior officers disagreed. They were under no illusion that the fortress would fall easily, and Kutuzov, among others, pointed out that the Russian bombardment had not been strong enough to cause serious damage to the Ottoman defences. Russian artillery could not shatter the thick earthen ramparts, and neither could it destroy the Ottoman cannon that remained silent in expectation of the assault.[23] Even more concerning was Prozorovskii's decision to commit only a small part of

---

19 Entries for April 19–21 in the Journal of Military Operations of the Main Corps in *M. I. Kutuzov: Sbornik dokumentov*, vol.III, pp.151–152.
20 See the Journal of Military Operations of the Main Corps and Kutuzov's orders in *M. I. Kutuzov: Sbornik dokumentov*, vol.III, pp.147–159; Alexander Langeron, 'Zapiski grafa Lanzherona. Voina s Turtsiei 1806–1812 gg', *Russkaya starina* 4 (1908), pp.229–230.
21 'Trenches have been completed everywhere; the heavy and siege batteries have been constructed and the bombardment has commenced; the few remaining works are nearing completion,' read the entry in the Journal of Military Operations of the Main Corps for April 28. *M. I. Kutuzov: Sbornik dokumentov*, vol.III, p.166. Also see Prozorovskii to Emperor Alexander I, April 23/May 5, 1809, in *M. I. Kutuzov: Sbornik dokumentov*, vol.III, p.182.
22 Prozorovskii to Rodofinikin, May 20/June 1, 1809; Prozorovskii to Emperor Alexander I, April 23/May 5, 1809, in *M. I. Kutuzov: Sbornik dokumentov*, vol.III, pp.166, 182–183.
23 Langeron mentions this issue several times in his memoirs, including while discussing the assault on Brăila. Langeron, 'Zapiski', p.226. Also see interesting details in Ivan Paskevich's

the Main Corps for the assault – just 8,000 men were to assail a fortress defended by almost 12,000 Ottomans. Prozorovskii argued that he only intended to capture the outside entrenchment. According to the rules of siegecraft, if this was accomplished, the garrison would have to surrender.[24] Kutuzov and other officers were doubtful it would work and urged patience and systematic reduction of the fortress. A premature direct assault, especially with insufficient forces, risked heavy casualties and a defeat, he warned. Their advice was ignored.

On April 30, Prozorovskii approved the disposition for the storming of the outer part (retrenchment) of the Ottoman defences.[25] It was an unusual plan because it proposed a limited assault on the Ottoman fortress (upon which the commander in chief insisted) and contained a strict injunction for soldiers not to advance beyond the retrenchment or attempt storming the citadel itself. The disposition for attack specified that 'the column commanders must closely observe and sternly remind their men that upon scaling the [outer] earthwork, nobody must venture forward for more than ten steps; any violators would be treated as traitors.'[26] The plan called for three assault columns, under Major Generals Sergei Repninskii II, Nikolai Zakharovich Khitrovo, and Vasilii Vyazemskii, to storm the Ottoman positions from multiple directions. Each column comprised three infantry battalions, preceded by 60 volunteers (*okhotniki*) who were to spearhead the assault, 40 men with fascines and ladders to help in crossing the deep trench, and 30 engineers to defuse mines that the Russian scouts reported around the fortress. Infantry battalions and dragoon squadrons were deployed behind each column to serve as the reserves and sustain the momentum. Because the Ottoman outer defences formed an irregular oval, with the left side being closest to the citadel and its cannon, the Russians decided to target the right side of the fortress, which was farthest from the castle. Vyazemskii's leftmost column, however, was designed as a partial diversion – upon approaching the enemy position, the column was to split into two, with one part supporting the central assault column while some 500 men would draw 'along the Danube's bank' near the leftmost enemy position, shouting and firing muskets but not attempting to assault the enemy positions. Under the cover of darkness, these men were to generate as much noise as possible to divert the garrison's attention (and draw away troops) while Repninskii (on the right) and Khitrovo (in the centre) breached other sectors of the retrenchment.[27]

---

memoirs in Sherbatov, *General-fel'dmarshal Knyaz Paskevich, ego zhizn' i deyatel'nost* (St Petersburg: V. A. Berezovskii, 1888), pp.55–56.
24 Prozorovskii to Emperor Alexander I, April 23/May 5, 1809, in *M. I. Kutuzov: Sbornik dokumentov*, vol.III, p.183.
25 Krasovskii, 'Iz Zapisok', p.503.
26 Disposition for the Storming of the Brăila Retrenchment, n.d. [probably April 30, 1809], RGVIA, f. VUA, d. 2929, ll., pp.108–108b. The disposition has also been published in *M. I. Kutuzov: Sbornik dokumentov*, vol.III, pp.172–174. On the feint attack, see Kutuzov to Essen III, April 19/May 1, 1809, in *M. I. Kutuzov: Sbornik dokumentov*, vol.III, pp.175. Also see Petrov, *Voina Rossii s Turtsiej 1806–1812 gg.*, vol.II, pp.220–223.
27 Disposition for the Storming of the Brăila Retrenchment, n.d. [probably April 30, 1809], RGVIA, f. VUA, d. 2929, ll., pp.108–108b.

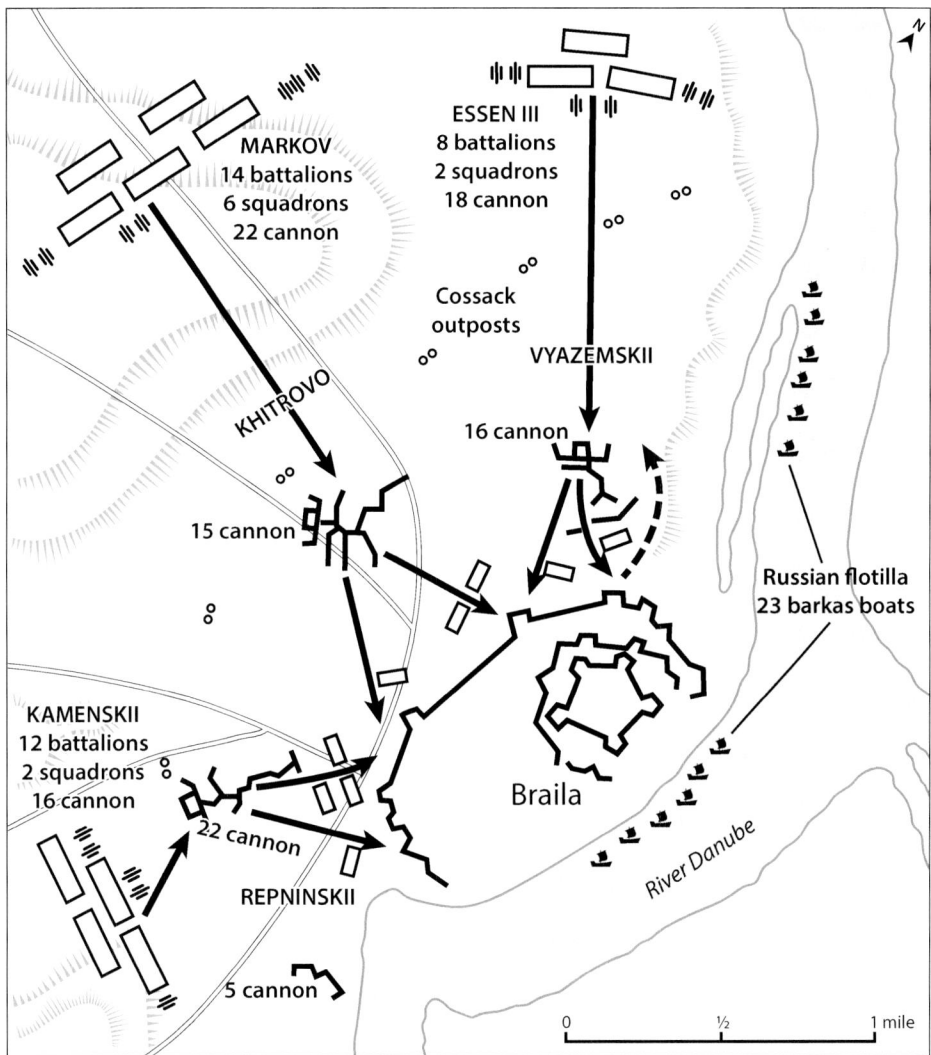

The assault on Brăila.

The plan of attack provoked much debate amongst the senior Russian officers, who doubted its efficacy. Harting (whom the soldiers nicknamed 'the Coffin Maker' for his earlier miscalculations) believed that the artillery bombardment had done its job of damaging the Ottoman defences.[28] Prozorovskii agreed and, as the commander in chief, insisted on the assault. 'Storming this fortress was a long-standing dream of the old man,' acknowledged General Langeron, who remembered the field marshal wishfully reciting examples of the celebrated Russian captures of Ochakov and Izmail. 'He told me about this in Jassy two months earlier, but I disagreed with him.

---

28 Afanasii Krasovskii, 'Iz zapisok general- adjutanta Krasovskago', in *Russkii Vestnik* 8 (1880), p.503.

The army did not approve of it either. Kutuzov dissented too, but [Prozorovskii] ordered it and everyone had to submit.' In the wake of Suvorov's bloody but celebrated storming of Izmail, Prozorovskii might have been truly 'yearning' to equal the great Russian generalissimo's exploit.[29] In his last order before the assault, the commander in chief exhorted column commanders to gallantly lead the assault and, 'with the Lord's help,' not only seize the Turkish entrenchment but to construct their own 'in a few minutes.' To reassure the no doubt disconcerted officers, Prozorovskii reminded them that he had served in the army for 60 years and had had a chance to witness 'everything that might transpire in war and have gained plenty of experience from it all.'[30] The prince's arrogance, overconfidence, and obstinacy were on clear display.

* * *

At 2:00 a.m. on May 2, three Russian columns left their camps and advanced toward the Turkish position.[31] An hour later, the last flare signalled the start of the assault. Almost immediately, things went downhill. 'The soldiers advanced still carrying their knapsacks,' bemoaned one officer. 'No one remembered to tell them to leave them behind, for the absurd and lethal *caporalisme* that infected the entire Russian army made their immediate commanders fearful of ordering them to take off knapsacks since the regulations required soldiers to carry them.'[32] Moving in darkness, the columns failed to coordinate their movements. While Repninskii and Khitrovo were just starting to march, Vyazemskii's men moved through the *vorstadt* (outside suburb), where an engineer leading the column made a wrong turn, leading it toward a deep ravine that the Russians mistook for the ditch in front of the retrenchment. The troops charged, yelling, as per orders, 'only to realise that it was just a gully,' wrote Ivan Paskevich, the future distinguished field marshal who in 1809 served as a staff officer in Prozorovskii's headquarters. 'So we climbed out of it and ran forward' toward the Ottoman positions. By this point, it was too late; the Turks had been forewarned. They waited until the leading Russian troops were crowded along the edge of the ditch before opening 'a fierce canister and musket fire' almost point-blank,

---

29 Langeron, 'Zapiski', pp.231–232. In his report to Emperor Alexander, Prozorovskii does state that Suvorov's success at Izmail encouraged him to assault Brăila. Prozorovskii to Emperor Alexander, April 23/May 5, 1809, in *M. I. Kutuzov: Sbornik dokumentov*, vol.III, p.185.
30 Prozorovskii's order to the Army, April 19/May 1, 1809, cited in Petrov, *Voina Rossii s Turtsiej 1806–1812 gg.*, vol.II, p.223.
31 The disposition did not indicate when the assault would commence, simply starting that 'the column commanders would be informed about the hour of attack.' However, Kutuzov's orders to Essen III, Sergei Kamenskii, and D. Akimov, who commanded the flotilla, specified that the first signal flares would be fired at 2:45 a.m. and the second ones 15 minutes later, 'and these flares should serve as the signal for assault' at 3:00 a.m. The sun rises around 5:50 a.m. at Brăila in May, so the plan would have given the Russian columns about two hours of darkness to initiate the assault and the rest of the morning to complete it. Kamenskii's after-action report confirms that his troops left the camp at 2:00 a.m., gathered at the rallying points, and 'upon the firing of the second flare' began the assault. See documents in *M. I. Kutuzov: Sbornik dokumentov*, vol.III, pp.175, 177–178.
32 Langeron, 'Zapiski', p.233.

killing and wounding dozens and dampening the attack momentum.[33] Meanwhile, on the extreme left, Vyazemskii's *okhotniki* pushed on, climbing across the moat and reaching the top of the Ottoman ramparts. But amid the darkness and confusion, Vyazemskii – 'very well educated and knowledgeable in military matters but rather careless and sluggish,' in the words of a fellow general – could not comprehend what was happening, so he did not rush the rest of his men to build upon the initial success, allowing the Turks to counterattack and cut down most of the *okhotniki*.[34]

The feint assault on the left flank failed to distract the defenders, who opened 'a vicious musket and artillery fire' on the main Russian assault columns. 'The canister fire claimed many of the ladder-carriers and some ladders never made it to the moat,' wrote General Sergei Kamenskii in his after-action report. Upon approaching the moat, some units discovered that instead of having been given seven ladders, there were just two or three, and even those had many rungs broken. The soldiers poured into the 15-foot-deep trench but could not climb out of it, while the Turks 'mercilessly killed them with musket fire and hand grenades.'[35] There was utter confusion; Paskevich, wounded in the head and crawling through the crowded ditch, remembered hearing soldiers shout that they had been betrayed and that ladders were missing. The groaning and moaning drowned out officers' orders, while the wounded soldiers scuttled in the ditch and tried to hide from the fire being directed at them from the top of the rampart.[36]

Standing on open ground on the left flank, Kutuzov must have had flashes of his own horrifying ordeal at Izmail as he observed, amid the bursts of gunfire that illuminated the smoke-filled battlefield, Russian soldiers descending into the ditch but struggling to get out of it. Time and time again, officers would collect a group of men in the ditch and try scaling the glacis; the few brave men would reach the summit of the retrenchment, only to be cut down by the Ottoman bullet or sabre. Amid this mayhem, the soldiers hunkered down at the bottom of the ditch. They kept shooting their muskets 'into the air and at each other, wasting ammunition and inanely shouting "hurrah," which only exacerbated disorder.'[37] Major General

---

33 Excerpts from Paskevich's memoirs appeared in print in Major General Prince Alexander Sherbatov's *General-fel'dmarshal Knyaz Paskevich* (1888), a seven-volume magnum opus examining the war hero's career based on archival sources and his unpublished private papers. Later, Paskevich's memoirs of 1812 appeared in print – V. Kharkevich, ed., *1812 god v dnevnikakh, zapiskakh i vospominaniyakh sovremennikov* (Vilna: Shtab. Vilen. Voen. Okr., 1900, vol.I, pp.82–111); A. Tartakovskii, ed., *1812 god v vospominaniyakh sovremennikov* (Moscow: Nauka, 1985, pp.75–105) – but his earlier writings are still unpublished and remain preserved at the Russian State Historical Archive, fond 1018, opis' 9. The quote is from Sherbatov, *General-fel'dmarshal Knyaz Paskevich*, vol.I, p.61. Also see Kutuzov to Prozorovskii, May 1/13, 1809, in *M. I. Kutuzov: Sbornik dokumentov*, vol.III, p.198.
34 Langeron, 'Zapiski', p.233. After the assault, Vyazemsky was investigated for his failure to support the *okhotniki* in time, but he was ultimately acquitted. Kutuzov to Markov, April 23/May 5, 1809; Markov to Kutuzov, April 28/May 10, 1809, in *M. I. Kutuzov: Sbornik dokumentov*, vol.III, pp.181, 190–191.
35 Sergei Kamenskii to Kutuzov, April 22/May 4, 1809, in *M. I. Kutuzov: Sbornik dokumentov*, vol.III, p.178.
36 Sherbatov, *General-fel'dmarshal Knyaz Paskevich*, vol.I, p.61.
37 Langeron, 'Zapiski', p.235.

Repninskii was slightly more successful in his attack on the right wing – some of his men penetrated the outer Ottoman defences and even seized one of the bastions. But they were left unsupported and vulnerable to a counterattack. In the end, they were driven into a deep moat, 'from where they could not get out and where they died without being able to defend themselves.'[38] Belatedly Prozorovskii rushed in reinforcements, but the Russian assault columns, bogged down, could not make any progress and were mowed down by what one eyewitness described as 'the most ferocious enemy fire; one may even say, the canister fire rained down upon us as a hail of [lead].'[39]

The rising sun cast light on a gut-wrenching panorama of carnage and human folly. 'The glacis and the moat were covered with numerous corpses,' recalled an eyewitness.[40] Prince Prozorovskii spent that night on top of the hill on the right flank. He was confident of victory and kept sending out adjutants to learn what was going on in the columns. No one dared tell him that it was time to beat a retreat. When the wounded Paskevich managed to get to him and explain how dire the situation was, 'the field marshal refused to believe my words,' Paskevich remembered. Others thought that he was delirious from his wound. 'Such comments enraged me and unloaded my anger explaining that the third column was not simply repulsed but rather annihilated.' The doleful officer then lost his conscience and was carried away to the field hospital. While Prozorovskii equivocated, 'the slaughter of our men continued in the retrenchment.' Kutuzov, who had realised that the attack had failed, came in person to see the field marshal and told him flatly that it was time to retreat. The news threw the field marshal into despondency – he fell to his knees and began to cry and bewail the setback. Some remembered that Kutuzov stood by, calmly observing the old man's despondency, and then telling him brusquely, 'You need to get to grips with it, for worse things have happened. I lost the battle of Austerlitz that decided the fate of Europe, and still I did not cry.'[41]

Around nine o'clock in the morning, Prozorovskii ordered the troops to fall back. The surviving Russian soldiers came to rue the daylight, for their retreating columns served as perfect targets for the Ottoman counterfire. In an incredible display of steadfastness and gallantry, the soldiers climbed out of the moat, straightened their lines despite the murderous fire that claimed so many of them, and then moved back while still being pummelled by the elated Turks.[42] The loss of life in this 'accidental, so as not to say needless or even senseless, assault,' as one participant ruefully put it, was startling.[43] Officially, the attackers lost 2,229 killed and 2,550 wounded, though

---

38 Langeron, 'Zapiski', pp.235–236.
39 Sergei Kamenskii to Kutuzov, April 22/May 4, 1809, in *M. I. Kutuzov: Sbornik dokumentov*, vol.III, p.178.
40 Sherbatov, *General-fel'dmarshal Knyaz Paskevich*, vol.I, p.61.
41 Langeron, 'Zapiski', p.237. Langeron thought Kutuzov's comment was 'rather frivolous.' Also see Krasovskii, 'Iz zapisok', p.504; Mikhailovskii-Danilevskii, *Opisanie Turetskoi voiny s 1806 do 1812 goda*, p.112; Petrov, *Voina Rossii s Turtsiej 1806–1812 gg.*, vol.II, p.228.
42 Prozorovskii to Emperor Alexander, April 23/May 5, 1809, in *M. I. Kutuzov: Sbornik dokumentov*, vol.III, p.184.
43 Langeron, 'Zapiski', p.232.

privately it was said that the number was almost twice as high.[44] The 13th Jäger Regiment went into action with 1,100 men; six hours later, all but 200 of them were gone.[45] 'The splendid grenadier battalion of the Vyatskii Infantry Regiment was completely annihilated,' mourned a Russian general. 'All in all, two hundred officers were killed or mortally wounded.'[46] According to General Langeron, the Ottoman commandant of Brăila sent 'sacks filled with 8,000 Russian ears' to Istanbul.[47] For the Ottomans, Brăila stood as a counterpoint to Izmail.

* * *

The Brăila debacle effectively ended the Russian offensive of 1809 before it even got fully underway. Stunned and dejected, Prozorovskii halted all offensive movements and concentrated his efforts on a protracted blockade of Brăila, just as Kutuzov and others suggested all along. For two weeks, the Main Corps stayed at the Ottoman fortress, firing thousands of projectiles that devastated the surrounding area.[48] At last, the prince summoned the council of war, which voted to lift the blockade on May 19 and take up new positions on the northern bank of the Danube.[49] 'Before departing, Prince Prozorovskii decided to seek vengeance on the Brăila garrison,' remembered one Russian general. 'He ordered all neighbouring villages to be burned to the ground and destroyed. The unfortunate inhabitants were allowed to keep only a couple of bulls and a small wagon for a few items.' In practice, the destruction of villages involved a large-scale looting, with senior officers 'shamelessly' participating in the plundering. According to Langeron, Prozorovskii remained blissfully ignorant of their involvement because 'no one around him dared to report it while the generals did not want to squeal on each other.'[50]

The setback at Brăila sapped whatever was left of Prozorovskii's initiative. Full of doubts and anxiety, the field marshal spent the next few weeks at the campsite, imagining threats and obstacles that, he argued, prevented him from engaging the Turks. One day, it was a joint Anglo-Turkish expedition that he thought was about to target the Russian Black Sea coast; the next, he argued that Austria was about to defeat Napoleon in the ongoing War of the Fifth Coalition and would then threaten Russia's western provinces. These were wild claims, unsupported by any facts, and reflective of the depth of the prince's paranoia. Even when the Serbs sent him frantic calls for help against the invading Ottoman army, Prozorovskii refused to provide any succour, effectively abandoning Russia's allies to Turkish vengeance. Nor did he

---

44   Petrov, *Voina Rossii s Turtsiej 1806–1812 gg.*, vol.II, p.228.
45   Krasovskii, 'Iz zapisok', p.503.
46   Langeron, 'Zapiski', p.236.
47   Langeron, 'Zapiski', p.237.
48   Kutuzov wrote lengthy instructions on how to conduct the siege works. *M. I. Kutuzov: Sbornik dokumentov*, vol.III, pp.180–181. Also see the Journal of the Military Operations of the Main Corps, in *M. I. Kutuzov: Sbornik dokumentov*, vol.III, pp.187–189.
49   Prozorovskii to Emperor Alexander, April 30/May 12, 1809, in *M. I. Kutuzov: Sbornik dokumentov*, vol.III, pp.195.
50   Langeron, 'Zapiski', pp.239–240. Langeron singles out Duty General Tyszkiewicz and his assistants who, in his words, gained 'scandalous fortunes' through plundering.

comply with Emperor Alexander's repeated orders to cross the Danube and coerce the sultan to accept the peace terms. 'Morally devastated and physically unable, [Prozorovskii] seemed afraid of Austria, Hungary, weather, in short, of everything,' wrote a nineteenth-century Russian military historian.[51]

By now, the news of the bloody setback at Brăila had reached St Petersburg, causing an uproar. The czar had expected news of a triumphant Russian crossing of the Danube, rather than of a rout with thousands of casualties, less than one month into the campaign. Prozorovskii rushed to justify his actions, and in doing so, he did his best to shift the responsibility onto the shoulders of his subordinates, whom he accused of incompetence, insubordination, and cowardice. His lengthy letter to Czar Alexander contained a litany of complaints and accusations against everyone and everything except the field marshal himself.[52] The czar carefully examined these explanations and responded with a special memorandum that outlined his own views on the siege warfare in the Danubian Principalities. He castigated the field marshal for launching a poorly thought-through assault on the Ottoman fortress: '

> Average measures never succeed in such challenging military operations, since their moderation and caution impede their own success. The objective should be worthy of such an extreme action. In the present case [of assault on Brăila], I do not see such an action warranted; therefore, if the objective is not worthy of resolute action, it is better not to undertake it at all.

Alexander distinguished two types of siege assaults, 'an assault through the breaches, when the dense columns [of troops] can advance into the city' and 'an escalade', a storming that involved scaling defensive walls or ramparts with the aid of ladders and other equipment. He pointed out that the siege of Brăila would have benefited from the former approach – a methodical reduction of the defences to create a breach and then storm the fortress. 'An unsupported assault requires exceptional resources and usually sacrifices a great number of men, even when it is successful; it also produces so much confusion [and losses] that even victory cannot fully justify it.' The czar reserved separate criticism for Prozorovskii's operational plan that focused on capturing enemy fortresses before crossing the Danube. 'By besieging fortresses and adopting a defensive posture, we stand to lose one of the most important factors in this war, which is the [Ottoman] fear of the deep Russian advance into their territories,' commented Alexander. 'If contrary to the [Ottoman] expectations, our army had advanced across the Danube, its movement would have certainly caused panic in Constantinople, engulfed as it is by disturbances and revolts. This panic alone would have been worthy of a great victory.'[53]

---

51 Petrov, *Voina Rossii s Turtsiej 1806–1812 gg.*, vol.II, pp.233–310.
52 He blamed the 'inexperience of the generals that engendered indecision' and accused his officers of failing to 'show sufficient courage and gallantry,' while the troops did not 'demonstrate sufficient subordination and confidence in their commanders.' Prozorovskii to Emperor Alexander, April 23/May 5, 1809, in *M. I. Kutuzov: Sbornik dokumentov*, vol.III, p.184; Petrov, *Voina Rossii s Turtsiej 1806–1812 gg.*, vol.II, p.229.
53 Printed in Mikhailovskii-Danilevskii, *Opisanie Turetskoi voiny s 1806 do 1812 goda*, pp.115–116.

Prozorovskii did not live long enough to implement a new strategy; exhausted and demoralised, he died in August 1809. A month later, his successor, General Peter Bagration, launched the first trans-Danubian offensive. He deliberately avoided costly sieges and assaults and instead emphasised the importance of methodical reduction of Ottoman fortresses and, most crucially, targeting the enemy field army. Such an approach did produce results – Bagration blockaded the fortress of Măcin on 14/26 August, completed construction of artillery batteries two days later and commenced the bombardment that forced the garrison of 338 Ottoman soldiers to surrender on 18/30 August. This success was then replicated at Hirsova (surrendered on 3 September), Constanta (9 September), Mangalia (15 September), and Izmail (26 September), all of which fell without the Russians resorting to assaults. By October 1809, Bagration blockaded Silistra, one of the strongest Ottoman strongholds in the Danube valley, forcing Ottoman Grand Vizier Yussuf to halt his invasion of Serbia and redirect some 50,000 men to the lower Danube. Bagration met the vizier's advance head-on, fighting a numerically superior Ottoman army to a draw at Tataritsa in October.

These victories raised prospects of a quick end to the war. However, the perennial logistical challenge of sustaining an army so far from its bases forced the Russian commander to lift the blockade of Silistra in late October and return to the left bank of the Danube; his last success actually accomplished what Prozorovsky could not – the fortress of Brăila capitulated on 3 December, after almost three-week long blockade. Though Bagration accomplished more in four months than Prozorovskii had in over a year, the czar disapproved of his conduct of the campaign and complained that Bagration's decision to withdraw had effectively forsaken all the gains. Ignoring the facts on the ground, Alexander insisted on a more forceful prosecution of the war, which Bagration found an impossible task. In the ensuing quarrel with the czar, the general was forced to resign his command.

The new commander of the Army of Moldavia was General Nikolay Kamenskii, young, headstrong, and impetuous, who ignored both the dire lessons Prozorovsky had learned at Brăila and the success Bagration had gained in gradually reducing Ottoman fortresses. Instead, Kamenskii insisted on more forceful and vigorous prosecution of the war and, on 22 July/3 August 1810, without making the required preparations, ordered the storming of the Ottoman fortress of Ruse. The results were devastating as the Russian troops discovered, once again, that their ladders were too short to climb out of the Ottoman moat and were caught in the enemy crossfire that decimated them. The failed assault claimed the lives of 78 officers and over 3,000 soldiers, while another 285 officers and 5,077 soldiers were wounded, most of them severely; this amounted to almost half of the 17,300 men that Kamenskii had committed to the assault. The hard lessons of the 1809 assault on Brăila that could have saved many lives were simply ignored.[54] Poignantly, a month after the failed storming, Kamenskii routed the Ottoman field army at Batin, which prompted the

---

54 Petrov, *Voina Rossii s Turtsiej 1806–1812 gg.*, vol.III, pp.96–112. After the failed assault on Ruse, Kamenskii reverted to the blockading, which eventually produced results as the fortresses of Ruse and Giurgiu capitulated in late September 1810.

garrison of Ruse to surrender. 'This was a clear proof that was absolutely no need for that calamitous assault,' ruefully noted a Russian participant.[55]

Kamenskii died shortly thereafter and was replaced by Kutuzov, who had long advocated a more mobile warfare that eschewed sieges and assaults in favour of deep offensive thrusts into Ottoman territory. He saw no utility in storming the Ottoman fortresses and, like Bagration, believed that the Russian army was strong enough to blockade them while conducting a vigorous campaign against the Ottoman field armies. Over the next year and half, while deprived of the much-needed reinforcements that Russia had to redirect in preparation for the war against Napoleon,[56] Kutuzov demonstrated the hallmarks of what many now recognise as Napoleonic combat leadership – avoiding sieges and blockading fortresses, targeting enemy's field armies through speed, manoeuvrability, and control of central position, bluffing as often as bludgeoning the enemy. Such an approach met with a resounding success as Kutuzov routed the main Ottoman army at Slobozia-Ruse in 1811 and forced the Ottomans to sue for peace.

* * *

The Russo-Ottoman War of 1806–1812 demonstrates that even though the time was long past when sieges had been the central experience of warfare, we should not entirely discount their significance. Certainly, battles had supplanted sieges as the central preoccupation of military commanders of the Revolutionary and Napoleonic Wars, but fortresses did retain their importance and could serve as a crucial determinant in the course of campaigns. Emperor Joseph II's decision to dismantle most of the fortresses of the Austrian Netherlands in the 1780s had deprived the Austrian forces of the strategic anchor points to slow down the invading French forces during the Revolutionary Wars. By contrast, the Ottoman fortresses, repaired and improved, continued to serve as bulwarks upon which successive Russian invasions stumbled and repeatedly demonstrated their strategic and operational relevance.

In his recent re-examination of the British army, Huw J. Davies explored how the British officers learned from their defeats and how knowledge acquisition and exchange had played a crucial role in elevating the British military to the pinnacle of its military prowess; the lessons continued to be learned and shared through

---

55 Aleksei Scherbatov, *Moi vospominaniya*, edited by A. Shiryaeva (St Petersburg: Nestor-Istoriya, 2006), p.70.
56 If given the necessary reinforcements, Kutuzov envisioned three corps, each acting independently of one another, launching a deep trans-Danubian incursion that would move around the fortresses and seek to penetrate deep into the Ottoman realm. The first corps would secure the Bulgarian countryside and pin down the main Ottoman army near Şumnu. The second corps would intercept the main routes to prevent any reinforcements from reaching the grand vizier. The third corps, meanwhile, would advance along the Black Sea coast, cross the Balkan Mountains, and head directly for Adrianople, which, in Kutuzov's opinion, would force the Turks to sue for peace. In its scale and ambition, the plan was unlike any previously considered during this war. The czar refused to grant the forces needed for such an operation, so the plan was never carried out. It would be another two decades before Kutuzov's ideas would be put into action, with success, during a new Russo-Ottoman war in 1828–1829.

'informal knowledge networks' as the officers 'wandered' from one theatre of war to another.[57] The story of the Russian 'knowledge networks' still needs to be told, but the siege warfare during the Russo-Ottoman War of 1806–1812 might offer a good framework for it. It does offer a litany of missed opportunities to learn and adapt, as time and again, the Russian commanders ignored the dire lessons of the past and chose brute force (and instant results it seemingly offered) as a quick solution to their tactical and operational challenges. But there were also moments when prudence prevailed. Much depended on the character and mindset of an army commander. While Bagration and Kutuzov demonstrated the necessary understanding of war, Prozorovsky and Kamenskii both acted impetuously and carelessly. After the failed assault on Ruse, Kamenskii refused to accept any responsibility for the failure and instead poured scorn on his men. 'Soldiers,' he wrote in a proclamation, 'You are responsible for this setback and for so many lost comrades. Some of you acted bravely, but most of you fell prey to fear.'[58] The few officers who dared to voice their disapproval were either told to mind their business or promptly removed from the army, thus leaving the 'yes-men' like 'the Coffin Maker' Harting, who had clearly failed to learn from earlier experiences, in a position to advise the commander-in-chief. Prince Aleksei Scherbatov, an eyewitness who was grievously injured at Ruse, felt dumbfounded by thoughtlessness and 'a horrific and pointless squandering' of human lives. 'Why would anyone gamble on brute force and chance that which could be gained through skill and a little bit of patience,' he wondered.[59] This points to a more fundamental issue of the Russian military experiences, one that transcends the Napoleonic period. The lives of Russian soldiers were consistently seen as expendable in the pursuit of military objectives. Their commanding officers faced little, if any, consequences for their abysmal mistakes. Despite knowing of the costly blunder committed in August 1810, Emperor Alexander spoke effusively of Kamenskii, who retained his command until his death. 'We cannot but lament the lack of success at Ruse,' the czar told his general, 'but this setback should not be assigned much importance. The losses we have suffered on this occasion would be more than made up when a new division joins the army.'[60]

---

57 Huw J. Davies, *The Wandering Army: The Campaigns That Transformed the British Way of War* (New Haven: Yale University Press, 2022).
58 Order to the Army, 26 July/7 August 1812, cited in Petrov, *Voina Rossii s Turtsiej 1806–1812 gg.* vol.III, p.108.
59 Scherbatov, *Moi vospominaniya*, p.69.
60 Cited in Petrov, *Voina Rossii s Turtsiej 1806–1812 gg.* vol.III, p.113.

2

The Siege of Izmail, December 1790: Russian Military Culture in the Early Revolutionary Period

Eugene Miakinkov

Siege warfare in general receives scant attention in the history of the Revolutionary and Napoleonic Wars, a time of grandiose battles, brisk movements, and well-established tropes about 'living off the land.' By 1789, it seemed like sieges and storms of fortresses would fade away as an exotic spectre of *Ancien Régime* warfare. Yet, as chapters in this collection demonstrate, sieges continued to be a resilient and important part of military operations across the myriad theatres of war of the period. This chapter examines Russia's siege warfare capabilities at the very start of the revolutionary period. It will analyse the grisly siege of Izmail, in modern-day Ukraine, by the Russian Imperial army in December of 1790, during the opening years of the French Revolution. Izmail was a Turkish fortress that refused to capitulate and which had to be taken by a desperate assault that lasted 11 hours. It sent shockwaves throughout Russia and Europe and was the turning point of the Russo-Turkish War (1787–1791). The analysis below, based on Russian-language original sources, goes beyond existing narratives, and frames this historical episode in a new way, as a window into Russian military culture.

Until recently, the siege of Izmail was the province of specialists, but over the last 20 years, several books have brought it to the attention of a wider and more popular audience.[1] Unlike existing accounts, this chapter treats the siege as both an illustration

---

1 Simon Sebag Montefiore, *Prince of Princes: The Life of Potemkin* (New York: Thomas Dunne Books, 2001), Chapter 30; Miakinkov, Eugene. *War and Enlightenment in Russia: Military Culture in the Age of Catherine II* (Toronto: University of Toronto Press 2020), chapter 6; and most recently, Alexander Mikaberidze, *Kutuzov: A Life in War and Peace* (New York: Oxford University Press, 2022), chapter 5. One of the first treatments of the siege in the English language based on Russian documents was rendered by Philip Longworth, *The Art of Victory: The Life and Achievements of Field Marshal Suvorov, 1729–1800* (New York: Holt, Rinehart and Winston, 1966), chapter 5. Much earlier work from the time of the Russian empire is N.A. Orlov, *Shturm Izmaila Suvorovym v 1790 godu*. (St Petersburg: Sklad izdaniia u V.A. Berezovskago, 1890).

of Russian military culture and a military episode. In the context of this volume, the siege of Izmail adds at least three important aspects to the era of the Revolutionary and Napoleonic wars. First, the siege provides an important East-European dimension to siege warfare of the period that is rarely discussed. Second, the chapter argues that Izmail peels away layers of Russian military culture to reveal the nature of the force that would challenge the French military machine and eventually break its backbone. Third, Izmail demonstrated how the Russian approach to siege-craft relied on the coordination of field artillery and a lightning storm by infantry rather than the more traditional use of siege artillery and blockade or prolonged investiture.[2] This Russian approach to siege-craft in the early years of the Napoleonic Era was sustained by various values and norms that defined Russian military culture, namely tenacious ability to take casualties, even among senior officers, capable leadership, often by personal example, respect for merit and initiative in promotion and awards, and attention to moral and spiritual needs of the soldiers.

In Russia, siege warfare remains understudied compared to the histories of the army or the navy. While several recent works have appeared,[3] probably the most comprehensive and thorough texts date from the Soviet and Imperial periods.[4] In English, Christopher Duffy synthesised much of Russian scholarship and provided an overview of the Russian sieges in the eighteenth century, as did Bruce W. Menning and John Keep.[5] Thanks to these works, we can piece together a picture of the state of Russian artillery and military engineering at the start of the Napoleonic Wars, and of the Russian technical capabilities in siege warfare. At the start of the 1780s, the Russian Empire made significant progress in developing its artillery park. By the beginning of the nineteenth century, the Russian Empire had 15 to 17 cannon-forging factories and 70 factories that produced an assortment of munitions. For instance, between 1782 and 1786, Russia manufactured 1,142 cannons for field and regimental artillery, garrison defence, and siege purposes. This growth in artillery and siege capabilities was due to the intensive development of metallurgy in the Russian Empire in the second half of the eighteenth century. Likewise, the development of gunpowder also became significantly more effective due to constant experimentation with and updates of gunpowder formulae. All of this meant that by the start of the siege of Izmail in 1790, Russia had 178 pieces of siege artillery, 378 cannons of various calibres that were distributed among infantry regiments,

---

2   S.N. Ionin, *Russkaia artilleriia: ot Moskovskoi Rusi do nashikh dnei* (Moscow: Veche 2006), p.77.
3   See Ionin, *Russkaia artilleriia*, and I. S. Prochko, *Istoriia Razvitiia Artillerii: S Drevneishikh Vremen i Do Kontsa XIX Veka* (St Petersburg: Poligon, 1994).
4   For example, L. Beskrovnyi, *Russkaia armiia i flot v vosemnadtsatom veke* (Moscow: Voennoe Izdatelstvo, 1958), N. E. Brandenburg, *500-letie russkoi artillerii. 1389–1889* (St Petersburg: Tipografiia 'Artilleriiskogo zhurnala', 1889) and A. A. Nilus, *Istoriia meterial'noi chasti artillerii* (St Petersburg: Tipografiia p.P. Skoikina, 1904).
5   Christopher Duffy, *Russia's Military Way to the West: Origins and Nature of Russian Military Power, 1700–1800* (London: Routledge & Kegan Paul, 1981), John Keep, *Soldiers of the Tsar: Army and Society in Russia, 1462–1874* (Oxford: Clarendon Press, 1985), and Bruce W. Menning, 'Paul I and Catherine II's Military Legacy, 1762–1801', in Frederick W. Kagan, and Robin Higham (eds.), *The Military History of Tsarist Russia* (New York: Palgrave, 2002), pp.77–105.

and 244 pieces of field artillery. To support its siege operations, by the end of the eighteenth century, the Imperial Russian Army had an engineering corps of 1,587 specialists that included companies of sappers, pioneers, and bridge engineers.[6]

The Russo-Turkish War, 1787–1791

The war between the Russian and Ottoman Empires overlapped with the first salvos of the French Revolution and was the product of Russian territorial designs and crumbling Turkish authority in south-eastern Europe. The origins of the war lay in the desire of the Ottoman Empire to reverse previous losses of territory to the Russian Empire, especially the annexation of the Crimea in 1774. Since then, Russian naval power in the Black Sea had increased by the day, the Russian army had made slow but firm gains in the Caucasus, meddling in the Ottoman sphere of influence, and interfering in trade with Persia. In response to Russian pressure, over the next decade, the Turkish military was, in the words of Isabel de Madariaga, 'steadily rearming with the help of the French experts and it was now or never as far as they were concerned.'[7] By the late 1780s, the Porte was finally ready for a revanche, and after St Petersburg renounced its ultimatum to return the Crimea, war was declared in August 1787. The military campaign centred around a series of sieges in modern-day Ukraine, Moldavia, and Romania, and progressed from one setback to another as far as the Porte was concerned. In 1787, the Ottoman attempts to take the Russian Black Sea fortress of Kinburn failed spectacularly, while the next year the Russians successfully besieged and stormed the large and important citadels of Hotin and Ochakov in present-day Ukraine. In 1789, the Russian forces besieged and took the fortress of Benderi in present-day Moldavia. As one Turkish fortress fell after another, by 1790 the Russian Imperial army had reached the gates of Izmail, the last and the largest remaining Ottoman stronghold in the Black Sea region. The siege of Izmail was a major turning point in the war that shattered the Ottoman military, confirmed Imperial Russia's possession of the Crimea, and marked the Ottoman Empire as 'the sick man of Europe'.[8]

Curiously, many careers of future giants of the Napoleonic Era intertwined under the walls of Izmail. Aleksandr Suvorov, who supervised the siege, would go on to command the allied army during the War of the Second Coalition in Italy and the Alps in 1799, defeating one French general after another. Roger de Damas, who commanded a regiment at Izmail, would go on to fight Napoleon as the Marshal of the Neopolitan forces. Aleksandr de Langeron, a junior officer at Izmail, would be one of the commanders in the Russian army during the Battle of Austerlitz. Armand-Emmanuel de Vignerot du Plessis, the future duc de Richelieu, who commanded a company at Izmail, became friends with the Russian Emperor Alexander I, and

---

6   Beskrovnyi, *Russkaia armiia*, pp.324–326, 355–364.
7   Isabel de Madariaga, *Russia in the Age of Catherine the Great* (London: Phoenix Press, 2003), pp.393–395.
8   Virginia H. Aksan, *Ottoman Wars: An Empire Besieged 1700–1870* (New York: Longman/Pearson, 2007), pp.167–170.

later served as the Prime Minister of France after the restoration. Finally, Mikhail Kutuzov, the future vanquisher of Napoleon's *Grande Armée*, was one of Suvorov's students in the art of war and during the siege personally led one of the attacking columns. All of them cut their teeth on the walls of Izmail.

From Indecision to Preparation

When Russian forces arrived at the gates of Izmail in the autumn of 1790, the war with the Ottoman Empire had been raging for three years under the command of Prince Grigorii Potemkin , the illustrious favourite of the Russian Empress, Catherine II. Izmail stood in a natural amphitheatre on a bank of the Danube. The fortress was protected by seven bastions and by massive fortifications that stretched around the city for 12 kilometres. Roger de Damas, the future marshal, was one of many French officers in Russian service who found themselves under the walls of the fortress. He wrote that 'the surface of the fortifications being of earth it was impossible to make a breach; the guns merely crumbled the earth; the damage was repaired during the night; and so no progress was made.'[9] Around the walls ran a deep moat filled with water from the river. The four entrance gates were brimming with artillery. Its southern side was open to the Danube but was protected by a flotilla of 23 vessels and gun batteries on the fortress walls.[10] Even if the Russians managed to break through this ring of defences, inside Izmail was the old citadel, which itself would have to be besieged. As the Russians continued to advance from the north, the retreating Turkish troops began to trickle into Izmail, along with provisions and military supplies.[11] The garrison at Izmail was unusually large, amounting to some 35,000 soldiers. This was part of the overall Ottoman strategy, which relied on a network of fortresses in southeastern Europe, connected by lines of communication for mutual support. This enabled the Ottoman Empire to hold large swaths of land with relative efficiency for hundreds of years. The armies of the Sultan tried their best to rely on this support network of fortifications instead of facing their enemies in pitched battles, on the open field. As William Fuller has observed, this strategy gave the Russian Imperial Army 'no recourse but to siege.'[12] Since the Porte liked to hide its forces in fortresses, Izmail soon turned from a fortress with a garrison into an army with a fortress.

In October 1790, the Russians began collecting reports from spies and defectors to gauge the enemy's strength concealed behind the walls of Izmail.[13] It is difficult to find sources about the state of the Ottoman defences. Potemkin estimated that the fortress had around 100 cannons, but four days before the storm, Russian soldiers captured a Turkish deserter who had tried to escape from the fortress and

---

9   Roger de Damas, *Memoirs of the Comte Roger De Damas (1787–1806)* (London: Chapman and Hall, 1913), p.137.
10 Beskrovnyi, *Russkaia armiia*, p.558.
11 Sergei Mosolov, 'Zapiski', *Russkii Arhiv*, 1 (1905), p.138; Damas, *Memoirs*, p.136.
12 William C. Fuller, *Strategy and Power in Russia, 1600–1914* (New York: Free Press, 1992), p.149.
13 Russian State Archive of Early Acts (RGADA): f.1453, op.1. d. 20, l. 1–2ob.

who confirmed the Russian estimates in detail. According to the Turkish prisoner, the fortress was guarded by more than 120 cannons that ranged in calibre from 9- to 33-pounders, organised in 16 batteries. Each battery was guarded by at least five soldiers, day and night. The largest batteries were concentrated around the four gates of the fortress.[14] The Russian authorities now had a clearer picture of what they were up against, but as it turned out, even this information was inaccurate. The real number of captured artillery after the siege was 265.[15]

Among the volunteers searching for glory and rewards around Izmail was the 28-year-old Count Grigorii Chernyshev. Writing to his brother, Prince Sergei Golitsyn, in late November 1790, he described the army's situation in grim terms. 'My dear brother,' wrote Chernyshev, 'the start of the campaign has been most unfortunate; everyone is feeling down, nobody knows what to do, and the Turks are celebrating.'[16] By late November, the situation at Izmail had reached a critical impasse. The Turks clearly were not going to be dislodged by the siege or intimidated by Russian manoeuvres or the Russian navy, and Potemkin did not have the confidence to do what they all knew was the only way to take the fortress, which was to storm it. To solve the crisis that was festering under the walls of Izmail and to extricate themselves from a politically vulnerable situation, Potemkin's commanders decided to call a military council.

The institution of the military council had been established by Peter the Great at the beginning of the eighteenth century in the *Military Statute* and was an important part of Russian military culture. This tradition continued in the nineteenth century when, at the famous Council at Fili in September 1812, Kutuzov and his lieutenants decided to retreat and give up Moscow to Napoleon instead of giving another battle to the *Grande Armée*. At Izmail, the military council played a crucial role twice, both as a military decision-making body and as a forum for political and military leadership. On this occasion, the council sent its decision to Prince Potemkin on 26 November.[17] In it, the generals argued that it would be futile to undertake any further actions.[18] They wrote that a deserter from Izmail had confirmed that the garrison was large and had more than enough guns to defend itself, as well as plenty of ammunition. They wrote that the Russian forces lacked the necessary siege artillery to reduce the mighty Turkish fortifications. They also pointed out that winter was fast approaching and that the army needed time to reach its winter quarters, which were far away. They warned that even if the bombardment and assault were launched immediately, the attack was unlikely to succeed and would cost thousands of lives. Finally, citing military principles of siege warfare, the council concluded that a storming of the fortress was impractical and that the siege should be replaced with a blockade. The Turkish garrison had only six weeks' provisions left, and the

---

14   Orlov, *Shturm Izmaila*, p.135.
15   Beskrovnyi, *Russkaia armiia*, p.558.
16   Grigorii Chernyshev, 'Pisma vo vremia osady Izmaila 1790 goda. Ot grafa G. I. Chernysheva k kniaziu G. F. Golitsynu', *Russkii Arkhiv*, 3 (1871), p.388.
17   All dates are from Russian original sources that used the Julian calendar which is 13 days behind the modern Gregorian calendar.
18   Orlov, *Shturm Izmaila*, p.37.

army of Her Imperial Majesty needed only to wait for the infidels inside to succumb to hunger, cold, and dysentery. The military council, therefore, advised that the army be withdrawn to winter quarters, leaving only enough forces behind to conduct the blockade. While the letter was travelling to Potemkin, before he had a chance to read it and write a reply, a full retreat had begun.[19] Potemkin was under pressure to bring the Porte to the negotiating table, and to do that Izmail had to fall. At the same time, he probably realised that the technical challenge of storming Izmail was beyond his military skill, so he sent for the one man whom he thought could take the fortress.

On 25 November, Potemkin personally wrote two letters to the eccentric *General-anshef*[20] Count Aleksandr Suvorov. Suvorov came from the minor Russian nobility, which had benefited under the reign of Peter the Great. Suvorov's appearance did not lead observers to think that he would one day become a great military leader. Short, with small sloping shoulders, wiry, and sickly, Suvorov had more in common with Prince Eugene of Savoy than with tall, portly giants like Potemkin. His life coincided with six major wars, which brought him to the pinnacle of military fame by the end of the century.[21] He became a field marshal at the age of 64 and eventually a *generalissimo*, a rare and unprecedented rank much later adopted by Joseph Stalin. Suvorov famously shared meals with his soldiers, his radiant, sometimes crude, personality fostered an extraordinary relationship with his peasant-recruits, and his knack of coming up with folksy, military maxims created numerous aphorisms that survive to this day.

In the first letter, Potemkin informed Suvorov that all the forces around the city, including the navy, had been placed under his command, and that he was to depart immediately for Izmail. By that point, Potemkin had mustered 600 artillery pieces of every calibre from every corner of the Russian army, for what he knew would be one of the most important sieges in the history of the Russian Empire. He also sent part of the Russian Black Sea Fleet – the Leeman flotilla of 38 ships and 48 smaller boats – to assist the army by closing off the side of the fortress that abutted the Danube.[22] He then encouraged Suvorov to look for the weakest places in the defences, writing that 'I personally think that the side of the city that is open to the Danube is the weakest.' At the same time, Potemkin had doubts that even the renowned Suvorov could take the fortress, and he instructed him to make sure he could retreat, in case, 'God forbid,' his assault failed. In the second letter, written later that day, Potemkin warned Suvorov about the discord among the commanding generals, which had led to inactivity and retreat, and singled out two officers whom

---

19 Orlov, *Shturm Izmaila*, p.130.
20 This is an old rank of the Russian Imperial Army, general-in-chief, corresponding roughly to full general.
21 Suvorov participated in six wars: the Seven Years' War (1756–1763), the Polish Civil War (1768–1776), the First Turkish War (1768–1774), the Second Turkish War (1787–1792), the Second Polish War (1793–1794), and the War of the Second Coalition (1798–1800). The most accessible biography of this fascinating figure remains Philip Longworth, *The Art of Victory: The Life and Achievements of Field Marshal Suvorov, 1729–1800* (New York: Holt, Rinehart and Winston, 1966). Apparently, Napoleon once commented that Suvorov was an officer with 'the heart but not the head of a great soldier.' Longworth, *Art of Victory*, p.301.
22 Beskrovnyi, *Russkaia armiia*, p.558.

Suvorov could depend on. Setting aside patronage networks and family ties, the prince wrote frankly that Suvorov should rely on Kutuzov and de Ribas instead of his relatives Aleksandr Samoilov and Pavel Potemkin. As if to lift Suvorov's spirits, he reiterated that the fortress was not impregnable – 'there are [weak] places, as long as there is good leadership.'[23] Finally, Potemkin hastened to inform Suvorov about the coup d'état by his generals and the decision by the military council to retreat:

> [I] propose your Excellency to act here according to your best judgment, continuing the Izmail enterprise or dropping it. Your Excellency, being there and having your hands untied, should not, of course, miss any opportunity that will be beneficial for us and that will add to the glory of our arms. Only please hasten to update me about what measures you are taking and inform the above-mentioned generals about your orders.[24]

Potemkin's letters reflected important aspects of Imperial Russian military culture that would directly translate into its strength during the siege and beyond, when the time came to face the Napoleonic challenge. There were clear recommendations of aptitude based on merit – de Ribas and Kutuzov were singled out as especially formidable. There was also a clear emphasis on personal initiative – 'act here according to your best judgment' – which underscored professionalism and trust between superiors and subordinates. Potemkin 'untied' Suvorov's hands and gave him the opportunity to assess the situation on the spot. He was free to attack or retreat, as long as he kept Potemkin informed.

The news of Suvorov's imminent arrival began to spread around the camp like wildfire. 'We are waiting the arrival of Suvorov every minute,' wrote Chernyshev in excitement.[25] The situation, however, looked unfavourable. Suvorov had 30,000 troops to capture a fortress with a garrison of 35,000, while the basic principles of siege warfare dictate that the attacker needs at least a three-to-one superiority over the besieged. Even before the first storming ladder touched the walls of Izmail, the new commander had to address the falling morale among the soldiers and prepare them for a terrifying undertaking. He had to convince the officers that victory was possible. He had to formally reverse the decision of the military council, which had undermined the army's confidence by concluding that the assault was impossible and that retreat was inevitable. The task before Suvorov was so monumental that he wrote as much to Potemkin a week before the assault, confessing to the prince that he 'could not promise anything' and that, despite Potemkin's earlier assurances, to Suvorov's eye, the fortress 'had no weaknesses.'[26]

Suvorov began the preparations by personally designing and supervising drills for the assault. As Chernyshev wrote approvingly, Suvorov put on 'exercises for the rehearsal of the upcoming storm.'[27] He called for the construction of ramparts and

---

23 Orlov, *Shturm Izmaila*, pp.128–129.
24 Orlov, *Shturm Izmaila*, p.130.
25 Chernyshev, 'Pisma vo vremia', pp.399–400.
26 Suvorov to Potemkin, 3 December 1790, in Orlov, *Shturm Izmaila*, p.131.
27 Orlov, *Shturm Izmaila*, p.131

moats identical to those of the fortress and ordered soldiers and officers to climb them with ladders.[28] He built dummy enemy soldiers and ordered his troops to attack them with bayonets fixed. These preparations were carried out at a feverish pace.[29] In 10 days of preparation, Suvorov's forces made 40 ladders and 27,000 fascines. Officers endlessly reconnoitred those parts of the fortress they were assigned to attack.[30] By personally demonstrating the drills, the general set an example for the officers, animated the soldiers, and lent the whole affair a new sense of urgency and confidence. Soldiers had no time to think about the December cold and the dangers of the assault – they were too busy building mock fortifications and practising drills. The morale of the troops was not neglected, nor was their spiritual preparation. As de Damas observed in his memoir, on the evening before the assault, 'the troops received the general benediction, and had the whole night at their disposal for rest, or, if they wished it, for the exercise of their various religious observances.'[31] Thus, religious and psychological preparation went hand in hand with the drills.

In Imperial Russia, religion was indispensable to military culture. Many Russian commanders were genuinely religious men, but that was beside the point. To them, religion served a practical purpose. Besides being an organised collection of beliefs and a source of comfort, religion was a tool for reaching into the soldier's soul and for making it receptive to the military values of self-sacrifice and respect for authority. As Bruce Menning wrote, Suvorov recognised and reinforced religious and patriotic sentiments, trying to awaken them in his recruits to strengthen 'common identity and loyalty to shared values.'[32] Geoffrey Best remarked that Suvorov took religion to such 'a heady pitch' that it almost served as a brainwashing mechanism.[33]

Next, Suvorov had to summon all his skills of leadership and persuasion to reverse the mood in headquarters and convince the soldiers gathered there, who only two weeks earlier had been ready to drop everything and retreat, to stay and attack. To that end, two days before the storm, Suvorov called a military council of his own. He used the military council to help him impose his will on the army; he wanted to make his view the view of his subordinates; he wanted his decision to storm the fortress to become the decision of his commanders.[34] It would become a tradition followed by his protégé Kutuzov in 1812 at Fili.

Surrounded by his officers, young and old, Suvorov delivered a motivating speech with his typical dramatic flair:

> Twice have the Russians approached Izmail, and twice have they retreated; now, the third time, all we can do is either take the city, or perish ... Retreat

---

28 Aleksandr Petrushevskii, *Generalisimus kniaz Suvorov* (St Petersburg: Stasulevicha, 1884), vol.1, p.384.
29 Chernyshev, 'Pisma vo vremia', pp.401–402.
30 Beskrovnyi, *Russkaia armiia*, p.559.
31 Damas, *Memoirs*, pp.138–139.
32 Bruce W. Menning, 'Train Hard, Fight Easy: the Legacy of A. V. and His 'Art of Victory', *Air & Space Power Chronicles*, 1986, p.82.
33 Geoffrey Best, *War and Society in Revolutionary Europe, 1770–1870* (Leicester: Leicester University Press in association with Fontana Paperbacks, 1982), p.44.
34 Orlov, *Shturm Izmaila*, pp.46–47.

from Izmail could weaken the resolve of our troops and encourage the Turks and their allies. But if we conquer Izmail, who will dare to stand in our way? I have decided to take this fortress, or die under its walls.[35]

Brigadier Matvei Platov, the most junior commander in the military council, who therefore had the first say, cried 'Storm!' and other generals immediately joined him.[36]

The Storm

On 7 December, Suvorov wrote an eight-page instruction for his army, outlining the disposition of forces, describing his tactics, and providing general rules of engagement. He divided his forces into three corps. *General-maior* (major general) Iosif de Ribas (1749–1800), a Spaniard in Russian service and the founder of the city of Odessa in modern-day Ukraine, was placed in charge of the navy and was to oversee the landing at what Potemkin called the 'weakest place' in the fortress. The two wings of the besieging army were to be commanded by *General-poruchik* (lieutenant general) Pavel Potemkin (1743–1796), the relative of the great prince, and *General-maior* Aleksandr Samoilov (1744–1814). Each of the three commanders had his forces further subdivided into three columns. Suvorov wanted Izmail to be attacked from nine directions simultaneously. If the element of surprise was somewhat lost due to deserters who had informed the Turks of the impending assault, Suvorov still could confuse the defenders as to where the main blow would fall.[37] In the end, Suvorov agreed with Potemkin's initial observation that the Danube side of the fortress presented the fewest obstacles. The other Russian columns needed to prevent the Turks from concentrating their forces in one spot and force them to spread themselves around the entire perimeter of the fortress.[38] This way, the enemy would not know where to focus his forces for counter-attacks and de Ribas and his men would have the best chance of success. Suvorov concluded his instruction with a brief note reflecting the wider Enlightenment concern with humanity and the rules of war. The instruction read: 'Christians and unarmed are to be spared, and of course, the same applies to women and children.'[39] Despite Suvorov's preparations and plans, one foreign observer in Russian service noted how each of the three corps

---

35   Quoted in Orlov, *Shturm Izmaila*, pp.47–48.
36   The document was signed in order of reversed seniority by the 13 generals under Suvorov's command: Brigadier General Matvei Platov, Brigadier General Vasilli Orlov, Brigadier General Fedor Vestfalen, Major General Nikolai Arsenev, Major General Sergei Lvov, Major General Iosif De Ribas, Major General Lasii, Major General Il'ia Bezborodko, Major General Fedor Meknob, Major General Petr Tischev, Major General Mikhail Kutuzov, Lieutenant General Aleksandr Samoilov, and Lieutenant General Pavel Potemkin. Orlov, *Shturm Izmaila*, pp.50–51; Damas, *Memoirs*, p.137.
37   Orlov, *Shturm Izmaila*, p.60.
38   K. Osipov, *Alexander Suvorov* (New York: Hutchinson and Co., 1944), p.88.
39   Dispozitsiia A. V. Suvorova k shturmu Izmaila, December 1790, in G. p.Meshcheriakov, (ed.), *A. V. Suvorov* (Moscow: Voennoye Ministerstvo SSSR, 1949–53), vol.2, p.532.

was in a rather lax state of readiness, 'and would have been defeated if the enemy showed more initiative.' There were many shortcomings with how the assault had been organised; indeed, it could be described as a 'very frantic affair.'[40]

The day before the storm, the Russian batteries opened a powerful barrage on the fortress, and the Russian ships on the Danube joined in. At first, the Turks responded energetically, but by the end of the day, their guns had fallen silent, likely to conserve ammunition for the impending battle.[41] The storm began at 5:30 a.m., two hours before sunrise. While there are several accounts of what happened that day, the narratives are riddled with gaps and uncertainties. Some accounts were written years after the event. They are as fragmented and confusing as for any other battle, and replete with macabre scenes and human error. For example, Suvorov had miscalculated the width of the moat and the height of the walls in his drills. The ladders that had been prepared for scaling the walls were too short, and the soldiers had to tie them together under a hail of musket balls. Columns got lost along the way to their staging positions, and during the fighting, many soldiers died from friendly fire. In addition to official reports and correspondence, *Premier Maior* (First Major) Andrei Denisov (the future leader of Cossack armies), Sergei Mosolov (the future major general), and several Frenchman in Russian service (Roger de Damas, Aleksandr Lanzheron, and Armand de Richelieu), left accounts of what took place on that day. Together they depicted not only the brutality of the fighting that surrounded Izmail on 11 December 1790, but also the culture, values, ideals, and strengths and shortcomings of the Russian military that would soon face the hammer blows of Napoleon's *Grande Armée*.

Andrei Denisov was under the command of *Brigadier* Vasilii Orlov, who led a column against the Bender gates, one of the four main entrances to the fortress, and one of the best-fortified. The column began to gather around midnight, and when the rockets signalled the start of the attack, Denisov and his men rushed toward the moat in front of the gates, where many of them were killed or wounded. Denisov and his men began to climb the steep ladders they had brought with them and reached the top of the gun battery, but could not take it. They were thrown back, 'beaten and injured.' Denisov was deafened by a grenade that landed between his shoulders, twice stabbed by enemy bayonets, and struck on the head with an artillery ramrod, causing a concussion. Under this pressure, he and others were forced down from the walls.[42]

Wounded and confused, Denisov retreated behind the moat and wandered around with musket balls whizzing past him until he encountered another band of Cossacks, where he found his commander, Orlov.[43] Here he received the news that his brother and his two cousins had already been killed. *Brigadier* Orlov confessed that he was shaken by the failure of the Don Cossacks to take the walls of Izmail

---

40   Armand de Richelieu, 'Journal de mon voyage en Allemange', *Sbornik Imperatorskago Russkago Istoricheskago Obshchestva*, 54 (1886), p.191.
41   Ionin, *Russkaia artilleriia*, p.77.
42   Adrian Denisov, *Zapiski donskogo atamana* (St Petersburg: VIRD, 2000), p.49.
43   Denisov, *Zapiski*, p.50.

and asked Denisov to help him regroup.[44] They gathered the remaining Cossacks, and with sabre in hand, Denisov once again advanced on the Bender gates, which turned out to be a dangerous mistake. Caught in the passion of the moment, he led his men head-on against entrenched enemy fire that immediately felled many of his comrades. 'A Cossack from my own regiment, Kiselev, arrived just in time, grabbed me by the hand and with the help of others, took me to the side and showed me my error,' wrote Denisov. After regrouping yet again, the column once more attacked the battery they had previously been repulsed from, which now had fewer defenders. As Denisov recounted, 'the Cossacks climbed over with heroic valour, and finally overpowered however many Turks they found there.'[45]

While Denisov was fighting his way through the Bender gates, on the opposite side of the fortress, Sergei Mosolov, in a column commanded by *General-Maior* Fedor Meknob, was trying to climb the western ramparts.[46] Mosolov recounted a specific episode during the siege that highlighted tensions between favouritism and the conceptions of professionalism and merit in Russian military culture. The officer who was supposed to lead Mosolov's battalion was *Sekund-Maior* (second major) Abram Marchenko. Marchenko led his battalion in the wrong direction and then disappeared, which compelled Meknob to ask Mosolov to take over and lead the men forward. 'I reasoned that it is better to heed his request than wait for a formal order,' wrote Mosolov. But he also hinted to his commander that if he agreed to make this extra effort, he should merit extra recognition. If Mosolov is to be believed, Meknob replied that 'if we take the fortress and remain alive, you will be doubly-rewarded.'[47]

Mosolov's men crossed the deep moat and assaulted an enemy bastion. He lost 312 soldiers in the process and was himself wounded with a musket ball through the brow and temple, leaving him temporarily blind in his right eye. 'If a trumpeter had not grabbed me from the cannon bastion,' wrote Mosolov, 'the Turks would have chopped off my head there.' As soon as he regained consciousness from his wound, he realised that only three soldiers around him were still standing; the rest were dead or wounded. Like Denisov, Mosolov regrouped for another attack, and to encourage his officers he shouted that the Turks had abandoned the bastion, which was a lie. But the lie served its purpose, and soon he had enough officers and soldiers under his command for another desperate charge. Despite losing more men, and with blood 'streaming' from his temple, Mosolov pressed on until 'we shouted Hurrah, burst into the bastion and took it.' He was weakened from blood loss and had to lie down, but in the meantime, the battalion commander, Marchenko, who had mysteriously disappeared before the attack, and was nowhere to be seen during the storm, reappeared at the bastion. One by one, Russian columns managed to scale the fortress walls, but the enemy did not surrender, and fighting continued for four more hours.[48]

---

44   RGADA: f. 1453, op. 1, d. 15, l. 1ob.
45   Denisov, *Zapiski*, p.50.
46   Mosolov, *Zapiski*, p.138.
47   Mosolov, *Zapiski*, p.139.
48   Mosolov, *Zapiski*, pp.138–139.

Given the importance the Russian military culture attached to religion, Orthodox priests played an important role during the assault. There is at least one record of a regimental priest not just inspiring but leading a regiment of Russian soldiers during the assault. When an attacking column on the western wall of the fortress got bogged down after its commander was mortally wounded, a regimental priest raised a big cross with Jesus the Saviour over his head and threw himself at the Turks.[49] Several sources related how this episode inspired the soldiers to secure a foothold on the walls. The priest, waving the cross, had called out to the vacillating soldiers, 'Steady, brothers! Here is our commander!'[50]

The surviving Russian accounts reveal the values of Russian military culture at the start of the Revolutionary Wars, including leading by example, being prepared for personal sacrifice, and displaying initiative. Denisov set a personal example of leadership at the Bender gates, even if it was sometimes misguided by his enthusiasm, and Mosolov accepted responsibility beyond his rank at a moment's notice without a formal order. The soldiers, too, played a vital part in both narratives – an anonymous trumpeter saved Mosolov, and the Cossack Kiselev pulled Denisov away from danger. This suggested that good officers won the respect and attachment of their men.[51] Finally, God was indeed a Russian 'commander' and faith could inspire where human leadership floundered.

In addition to the Russian accounts, there are several by foreigners, who left their impressions of the siege in their memoirs. Armand de Richelieu, a royalist during the French Revolutionary Wars and the future Prime Minister of France, became a noble vagabond in Potemkin's army after the revolution. He left one of the most personal and thoughtful reflections on the assault. Like many other foreigners, he was under the command of De Ribas, and he was placed in charge of several battalions of infantry that had been ordered to land on the Danube side of the fortress. Richelieu's observations about the Russian officers and soldiers during the assault reveal the extent to which the success of the whole affair hung by a thread. At times, generals and officers had to plead with and threaten their soldiers to get them to advance on the enemy instead of just firing their muskets. Every battery, every tower, every gate, was taken only after heavy losses. Confusion was general, and some Russian soldiers panicked and 'lost their heads,' fleeing back to their own lines 'with faces marked by horror and desperation.' In one instance, Richelieu wrote, his soldiers retreated no less than 50 times. Only the fact that the Turks did not attempt a pursuit turned the Russian soldiers around. After Richelieu's men succeeded in taking a Turkish battery, he took out his purse, which he had carried with him into battle, and distributed its entire contents to the soldiers on the spot.[52] Once in the city, not a minute passed that he did not see somebody's throat being slashed by the Russian troops.[53]

49　Orlov, *Shturm Izmaila*, p.69.
50　Lev Engelgardt, *Zapiski L'va Nikolaevicha Engelgardta, 1766–1836* (Moscow: Izdanie Russkago Arkhiva, 1868), p.117; Damas, *Memoirs*, p.141.
51　I thank Richard Hall for this insight.
52　Richelieu, 'Journal de mon voyage en Allemange', p.181.
53　Richelieu, 'Journal de mon voyage en Allemange', pp.184–185.

Richelieu did not differentiate the Russians' conduct or their military culture from the wider European practices of the time. Referencing the torrent of violence in post-revolutionary France, he thought that Europeans could not call 'barbarity a peculiarly Russian character[istic].'[54] He wrote how he witnessed the growing rage of the soldiers and officers at the tenacious resistance offered by the besieged, and how this rage spilled over onto civilians and the remaining garrison once the walls of Izmail were finally breached. Richelieu concluded that neither the strict hierarchy within the Russian army nor even Potemkin himself could have saved even a single Turkish life that day.

Another detailed account of the assault was left by Roger de Damas, the future marshal. Along with Prince Charles de Ligne, Armand de Richelieu, Aleksandr Lanzheron, and many other French and Austrian officers, de Damas was under the command of de Ribas, approaching Izmail from the Danube side, with the Russian navy.[55] De Damas was in charge of a regiment of 2,000 Livonian light infantry. His account offers a view of the 'weakest part' of the fortress from the perspective of a bewildered Westerner. Eight minutes after the signal for attack was given, de Damas with his men crossed the river in boats to attack the underbelly of the fortress. In those eight minutes, he lost close to 60 officers and soldiers. Upon disembarking, he and his column made it to the top of the rampart, overpowered the defenders, and turned captured cannons 'upon the Turks in the fortress.'[56] He related what he saw unfolding around the city walls while he stood on the rampart:

> I was joined by the aide-de-camp of General Ribas, who begged me to hold firm as long as was possible, because none of General Souvarow's columns had as yet been able to descend from the ramparts, though several had succeeded in reaching the summit, after losing half their men and climbing from corpse to corpse. They could not, however, beat back the Turks, who defended themselves from the inner base of the parapet, without losing ground.[57]

The siege was bitterly contested, with the fortunes of both sides hanging in the balance. The key moment was when the Russian troops succeeded in descending from the ramparts and opened one of the gates from the inside. Suvorov was waiting for this moment and immediately ordered his cavalry to rush into the city, accompanied by field artillery.[58] Now the Russian forces could blast their way through the fortress, one house at a time, and secure a victory.

The siege of Izmail concluded with a sudden massacre. What precisely started it is unclear, but it appears that during the surrender of the *Seraskier*

---

54 Richelieu, 'Journal de mon voyage en Allemange', p.189.
55 Elena Polevshchikova, 'Frantsuzkie volontery v Izmaile: neopublikovannaia zapiska grafa Lanzherona,' *Deribasovskaia-Reshilevskaia: Literaturno-khudozhestvennyi, istoriko-kraevedcheskii illiustrirovannyi almanakh,* 29 (2007): p.11.
56 Damas, *Memoirs*, p.140.
57 Damas, *Memoirs*, p.140.
58 It is not clearly how many field artillery pieces Suvorov brought into the city or what was their calibre. Ionin, *Russkaia artilleriia*, p.77.

(commander-in-chief)[59], the Turkish commander, with 4,000 troops, one of his bodyguards cut down a Russian soldier who was probably reaching to take the Seraskier's weapons as a trophy.[60] Westerners would be the only participants who described the ensuing savagery. As de Damas wrote:

> …the most horrible carnage followed – the most unequalled butchery. Two hours were employed in a hand-to-hand fight … Every armed man was killed, defending himself to the last; and it is no exaggeration when I say that the gutters of the town were dyed with blood. Even women and children fell victims to the rage and revenge of the troops. No authority was strong enough to prevent it.[61]

Finding himself surrounded by dead and dying soldiers and having narrowly escaped death from stray musket fire, de Damas was at the point of mental and physical exhaustion. He found a cot in one of the few houses still standing in Izmail 'and slept upon it for nineteen consecutive hours without once awakening.'[62] Disorder and teamwork, merit and incompetence, brutal fighting and leadership by example were all powerful markers of Russian military culture in the accounts of the assault. The participants attempted to impose a clear narrative on the chaos that had engulfed them and their enemies. Denisov and Mosolov created sharply distinct characters. Mosolov was professional and brave; Major Marchenko was a coward. The Cossack Kiselev saved Denisov's life. The available accounts make it clear that training, indoctrination, and drills had had a powerful effect. While Russian units retreated on many occasions, none of the sources report a mutiny. As the commander of one of the columns, and the future leader of the Don Cossacks during Napoleonic Wars, *Brigadier* Matvei Platov, the commander of the Cossack forces, wrote to general Samoilov, he saw with his own eyes how 'the example of bravery by commanders who were always in front, inspired soldiers.'[63]

The siege was over, but within Izmail's walls, the carnage and pillaging were unrelenting. The records left by the foreigners expose the limits of Enlightenment humanitarianism in Russian military culture. 'There could not be any talk of saving the wounded [enemies], almost all of them were mercilessly finished off. There were prisoners, who after seeing such terrible slaughter, died of fear,' wrote Aleksandr Lanzheron, future commander in the battles of Austerlitz and Leipzig, and one of the founders of the resort town of Odessa, in modern Ukraine.[64] In addition to that, according to Mosolov, 'after the siege the count [Suvorov] permitted the lower ranks to take in the fortress whatever they found for three days.' So much loot was available that soldiers were filling their hats and caps full of coins.[65] Even so, Richelieu would

---

59 A rank in the old Ottoman Army equivalent to full general.
60 Longworth, *Art of Victory*, p.173.
61 Damas, *Memoirs*, p.140.
62 Damas, *Memoirs*, p.141.
63 RGADA: f.1453, op.1, d.16, l. 1.
64 Polevshchikova, 'Frantsuzkie volontery', p.11.
65 Mosolov, *Zapiski*, p.139; Orlov, *Shturm Izmaila*, pp.78–79. I could not find this order, which several authors allude to.

later marvel that 'despite the strongest indiscretion that reigned in the Russian forces that day, that evening everything was brought to order' under Kutuzov's leadership, who was appointed the commandant of the conquered fortress.[66]

Casualties and Rewards

As the cries of the wounded subsided, de Damas provided one of the first assessments – a surprisingly accurate one – of what the assault had cost the Russians: 'Nine thousand Russians were killed and wounded, including several generals.'[67] The actual number was close to 10,000, with 400 out of 650 officers dead or wounded.[68] Lanzheron, who had fought his way into the city from the Danube side, noted that 'almost all of the columns lost a third of their soldiers.'[69] Almost all of the officers leading the nine attacking columns were killed or wounded. The Russian death toll in senior officers was unprecedented: 11 major generals, one brigadier, six colonels and more than 40 lieutenant colonels laid down their lives during the assault.[70] The military culture of the Russian Imperial army drove its members to extraordinary exertions, and Russian actions during the siege conformed to the masculine ideals of the Enlightenment, such as bravery, initiative, and self-sacrifice. *General-Maior* Boris Lassi was wounded in the hand but continued to fight until victory was declared. *General-Maior* Meknob suffered a severe leg wound and had to release his command to one of his subordinates. He died a few days later. *General-Maior* Sergei Lvov was also forced to give up his command after being wounded. *General-Maior* Count Ilia Bezborodko similarly gave up his command, but only just before he fainted from his wound. *Polkovnik* (colonel) Prince Dmitrii Lobanov-Rostovskii, who volunteered to participate in the assault and led 150 musketeers, was seriously wounded. *Sekund-Maior* Prince Drutskoi-Sokolinskii died while climbing the siege-works and was replaced by *Sekund-Maior* Prince Trubetskoi.[71] Dead princes, generals, and common soldiers, along with grand viziers, pashas, and janissaries, littered the walls, the trenches, and the streets of Izmail. As dishevelled Kutuzov wrote to his wife after the storm, 'there are so many things to do that I cannot even take care of the wounded; I need to restore order in the city, but there are more than 15 thousand dead Turkish bodies alone ... I can't even collect my corps together, there are almost no officers left alive.'[72]

The Russians gave their own dead proper burial in accordance with Orthodox custom, but they did not have the manpower, or the will, to provide similar for their fallen enemies. The surviving 10,000 prisoners were employed to clear Izmail

---

66 Orlov, *Shturm Izmaila*, p.191.
67 Damas, *Memoirs*, p.141.
68 Montefiore, *Prince of Princes*, p.580.
69 Polevshchikova, 'Frantsuzkie volontery', p.10.
70 Polevshchikova, 'Frantsuzkie volontery', p.10; Orlov, *Shturm Izmaila*, p.81.
71 'Suvorov to Potemkin, 21 December 1790', in Meshcheriakov, (ed.), *Suvorov*, vol.II, pp.551–561.
72 Mikhail Kutuzov, 'Arkhiv Kniazia M. I. Golensheva-Kutuzova-Smolenskogo, 1745–1813,' *Russkaia starina*, 2 (1870), pp.500–501.

of dead Turkish soldiers, civilians, and horses. It took six days to collect more than 25,000 corpses, and because the earth was by then frozen, they were all thrown into the Danube.[73] The atmosphere in the Russian camp was equally grim, with the surviving soldiers and officers anxiously searching for their friends and relatives. Among the officers looking for information about fallen relatives was Andrei Denisov. While he was capturing the bastion, he had learned that his brother and his cousin were not dead after all, but instead severely wounded.

> I found my brother half-dead,' he wrote, 'the bone in his arm above the elbow was entirely shattered, a musket ball has ripped through his entire foot, lodging itself in the big toe, that was extracted in my presence … In the same tent lay General Meknob, [and] my cousin, who was heavily wounded by two musket balls, and several of our regimental commanders and officers.'[74]

The day after the assault, while the exhausted soldiers and officers were recovering from the brutal fighting, the Russian army held a large prayer service under the thunder of cannons, again mixing religious with military ceremony. The service was led by the same priest who had heroically led a group of Russian soldiers during the assault.[75] As Suvorov informed Potemkin, the service was held next to a mosque that had been converted into the new Church of St Spiridion, the patron saint of miracles, since the capture of Izmail had taken place on his day.[76] This symbolic gesture legitimised the role of providence in Russian military culture, as well as the faith of its soldiers and officers in the divine saviour.

Ten days after the assault, Suvorov finished a 43-page report with the details of the battle and a long list of recommendations, which he sent to Potemkin. However, the report contained no details of the massacre. What reward awaited the brave priest who had spurred on the Russian soldiers at the moment of crisis? Did Meknob keep his promise to Mosolov? And what of the many soldiers and officers like Denisov and Damas who had survived the horrors of the walls and streets of Izmail?

Suvorov's report, which he based on letters sent to him by the commanders of the nine attacking columns, showed how even during the chaotic and confusing circumstances of a siege, Russian military culture strived to document the merit of all those involved. Junior officers had their eyes on their subordinates, while senior officers keenly observed their juniors for evidence of bravery, intelligence, or initiative that could earn them an award or a promotion. For his bravery, the priest received a gold ring in the shape of a cross, on a sash of St George.[77] The Order of St George had been established in 1769 by Catherine herself as the highest military honour in the Russian Empire. It had four classes, the first being the highest.[78]

---

73 Petrushevskii, *Generalisimus kniaz Suvorov*, vol.1, p.395.
74 Denisov, *Zapiski*, p.51.
75 Orlov, *Shturm Izmaila*, p.80.
76 'Suvorov to Potemkin, 13 December 1790,' in Meshcheriakov (ed.), *Suvorov*, p.540.
77 Engelgardt, *Zapiski*, p.117.
78 *Polnoe Sobranie Zakonov Rossiiskoi Imperii*, vol.18, no. 13387.

Despite being dishevelled and depressed, Kutuzov 'was the first to enter the city.' So boasted his wife, Katerina, in a letter to a relative. He received the order of St George, Third Class, and was named the commandant of the fortress.[79]

The young Count Chernyshev had played an important role and especially distinguished himself:

> I cannot leave behind, and not justly attest and recommend to your serenity ... her Imperial Majesty's Chamberlain Count Chernyshev, who was appointed by me due to his abilities and knowledge, to observe the actions of all the columns, who threw himself in all the dangers, fearlessly taking notes for the composition of this report, and who was employed by me on numerous occasions in many parts of the army with different assignments and is worthy of my special attention and request for rewarding him for courage and skill.[80]

Chernyshev was Suvorov's eyes and ears during the storm and provided the commander-in-chief and his staff with information the latter required in the moment. He reported back to headquarters about the progress of the attack and about the conduct of individual officers and soldiers. It was in part due to people like Chernyshev that commanders could cross-reference the accounts of the participants and letters of recommendation from field commanders and compose accurate reports.

Sergei Mosolov received even more praise:

> *General-Maior* Meknob attested to the brave spirit of those in his column during the storm: ... *Podpolkovnik* [lieutenant colonel] and cavalier Fedor Meller and ... *premier maior* Sergei Mosolov; these two climbed the bastion, Meller from the left and Mosolov from the right, and courageously drawing others to follow them, were wounded from the embrasures, the first in the neck, and the second in the head; and Mosolov, overcoming the heavy wound, turned to his men and served as an example to his subordinates in routing the Turks, who he took prisoner, and then continued to fight on, eliminating the enemy with superb valour, but also taking measures to protect the wounded.[81]

Though he would die the following day, Meknob faithfully noted Mosolov's persistence during the assault, as he had promised. Mosolov's sacrifice had been documented, his leadership had been praised, and the outcome of his initiative – the defeat of the enemy soldiers in the bastion and the assistance he provided to wounded comrades – had been extolled. But despite the strong recommendation

---

79 'Ekaterina Kutuzova to Aleksei Kutuzov, 6 January, 1791', in Ia. L. Barskov (ed.), *Perepiska Moskovskikh masonov XVIII veka* (St Petersburg: Izd. Otd-nie russkago iazyka i slovesnosti Imperatorskoi akademii nauk, 1915), p.77; Kutuzov, p.2.
80 'Suvorov to Potemkin, 21 December 1790', in Meshcheriakov (ed.), *Suvorov*, p.554.
81 Meshcheriakov (ed.), *Suvorov*, p.559.

he received from his dying general, Mosolov would write bitterly about his experience at Izmail. 'All of my extra efforts for the fatherland,' he complained, 'were for naught; for general Meknob died, and I received only the cross of St George, 4th class, along with another major, Shekhovskoi … who lost his arm.' What Mosolov really wanted was a promotion.[82]

Adrian Denisov, along with other majors, 'acted with courage' and set 'examples for others.' Moreover, the report continued, '*premier-maior* Golin, Denisov, *Polkovnik* [colonel] Petr Denisov, and *sekun-maior* Ivan Grekov and *kapitan* Ivan Karpov, overcame stiff resistance and hand-to-hand combat, and were examples for others to ascend the parapet.'[83] In conclusion, 'the volunteers and regiment of Denisov climbed the curtain wall using ladders with haste and courage, the valour of which is commendable.'[84] For his efforts during the assault, Denisov received the coveted Order of St George, Fourth Class. The events at Izmail reflected the vestiges of favouritism, and existing jealousies and bitterness, but also the honest and professional recognition of merit and ability. Chernyshev and his team of staff officers diligently recorded everyone's conduct; Mosolov and Denisov fought on despite being pushed back; many senior commanders fought until they passed out from their wounds or sheer exhaustion; Meknob did his duty of attesting to the conduct of his officers and soldiers before dying.

The French émigrés, who had exchanged the ravages of the revolution in France for the slightly more comfortable surroundings of Potemkin's army under the walls of Izmail, were rewarded as generously as the Russian officers, and no distinction was made between foreign and Russian participants in the siege. The actions of de Damas were carefully documented and reported to the commander-in-chief, Prince Potemkin. After landing with his surviving soldiers, he demonstrated 'such courage and zeal that despite the heavy enemy resistance, he passed through the line of fire, cleared the shoreline, pushed back the enemy, and continued to hold his position.'[85] *Polkovnik* Prince Charles de Ligne supervised the construction of Russian batteries under enemy fire and, during the assault, was wounded in the leg.[86] Empress Catherine wrote a personal letter to de Ligne, rewarding him with the Order of St George. The prince was among those who 'shared the dangers of mounting, without an open trench, without a battered breach, the formidable fortress of Izmail, where a whole army of enemies to Christian men were awaiting you,' wrote the Empress. 'The order of Saint-George,' she continued, 'having for the basis of its statutes the laws of honour and valour, – precious synonyms to heroic ears – is always by its institution eager to count among its valiant knights whoever gives proof of those military virtues.'[87] The case of de Ribas presents another clear example of how merit was documented, decided, and awarded. A month before

---

82   Mosolov, *Zapiski*, p.140.
83   'Suvorov to Potemkin, 21 December 1790,' in Meshcheriakov (ed.), *Suvorov*, pp.562–563.
84   Meshcheriakov (ed.), *Suvorov*, p.564.
85   Meshcheriakov (ed.), *Suvorov*, p.553.
86   Meshcheriakov (ed.), *Suvorov*, p.554.
87   'Catherine to Prince De Ligne, January 1790,' in Charles Joseph De Ligne, *The Prince De Ligne: His Memoirs, Letters and Papers* (Boston: Hardy, Pratt & Co, 1899), p.194.

the storm, on 16 November, when Potemkin was updating the Empress on the situation around Izmail, he wrote of de Ribas's success in eliminating the Turkish fleet and clearing the Danube of the enemy navy. 'Mentioning him I cannot pass over in silence his unparalleled zeal,' concluded the prince. A few days before the storm of the fortress, on 3 December, Potemkin again wrote that 'General-Maior Ribas deserves a lot, and even more good progress can be expected from him in the future.' On 20 December, Catherine wrote back to Potemkin that 'for Major-General Ribas on the first occasion I am sending the order of St George, 2nd class, which he has rightly earned, and then I leave it up to you how to continue to reward him.'[88] Finally, the humble soldiers, the true conquerors of Izmail, were ostensibly given permission to plunder the city as a reward for their tenacity and sacrifice. In addition to a huge amount of loot, all soldiers who participated in the siege received silver medals, which was customary in the Russian military in the eighteenth century.[89]

The question of how to reward Suvorov was much more complicated. In a letter to Catherine, Potemkin expressed his thoughts about how to recognise the mastermind of the siege. First, he recommended minting a special medal in Suvorov's honour in recognition of his service in taking the fortress. Second, he noted that among all the senior generals, Suvorov was the only one who had seen real action. With all of this in mind, wrote the prince, 'will it not please you to distinguish him with a rank of a lieutenant-colonel in the guards or as an adjutant-general.'[90] As Potemkin had suggested, the hero of the day was rewarded with a special medal and the rank of lieutenant colonel in the Semenovskii Guards – two highly symbolic, personal, and visibly distinguishing gestures. Minting a medal in honour of Suvorov's victory was a much more exclusive recognition of the unique significance of the event than any other award would have been. The medal, unlike an award, was a mark of individual merit, and it bore Suvorov's silhouette in the style of ancient heroes, thus conforming to the neoclassical style of the century. The medal carried historical significance and was Potemkin's and Catherine's way of showing Suvorov, the military, and the Russian public who 'owned' the victory at Izmail. The honour of a personalised medal was reserved for a select few. In recent times, only Potemkin had received similar recognition for his victory at the siege of Ochakov. Making Suvorov a lieutenant colonel of the Semenovskii Guards, of which the Empress was the colonel, was similarly a rare and significant honour. Suvorov shared that honour with only 10 other people in the Russian Empire, including Potemkin himself.[91]

---

88 Quoted in Nikolai Skritskii, *Georgievskie kavalery pod Andreevskim flagom. Russkie admiraly – kavalery ordena Sviatogo Georgiia I i II stepenei* (Moscow: ZAO Tsentrpoligraf, 2002), p.368.
89 Orlov, *Shturm Izmaila*, p.85. For instance, in 1788 participants in the siege of Ochakov received a specially minted medal celebrating their victory.
90 'Potemkin to Catherine, 24 March 1791,' in Viachislav Lopatin (ed.), *Ekaterina II i G. A. Potemkin: lichnaia perepiska, 1769–1791* (Moscow: Nauka, 1997), p.454.
91 Viachislav Lopatin, *Potemkin i Suvorov* (Moscow: Nauka, 1992), pp.216–217.

Conclusion: Russian Military Culture at the Dawn of the Napoleonic Era

This chapter presents a new interpretation of a pivotal historical moment in Russian history and advances a novel argument about the nature of Russian military culture, grounded in original sources. The siege of Izmail showcased Russian technical capabilities in siege warfare. Izmail demonstrated a way to take a major fortress while circumventing a long and expensive blockade or investment by employing field artillery and attacking columns. In this context, the preparations, the battle, and its aftermath reflected the broader contours of Russian military culture at the dawn of the Napoleonic era. Merit, training, professionalism, indoctrination, and religion all came together at the siege of Izmail. The willingness to absorb losses in soldiers and officers, leadership by example, observation and celebration of merit, and attention to morale elements in training were all crucial parts of Russian siege craft. Potemkin had given Suvorov the freedom to act independently and constantly informed him of the political situation. Upholding the principles of merit, he identified talent during the preparations for the siege based on distinction and past performance, and he recommended reliable officers to his chosen commander. Suvorov, for his part, in the best traditions of *Ancien Régime* professionalism and leadership, drilled his soldiers, sometimes in person, inspired his officers, and made time for religious ceremonies. Reflecting the larger values and ideals of late eighteenth-century Russian military culture, the battle that followed displayed many acts of leadership by example, episodes of personal bravery and sacrifice that were contrasted with cowardice, and instances of initiative and cooperation between soldiers and officers. Finally, both during the fighting and afterwards, when accurate attestations were being crafted, the honour and importance of attesting to valour played a significant role in the world of the Russian army, serving as a motivating factor and as part of officers' duty.

The aftermath brought scenes that must have troubled the officers and the commander-in-chief himself.[92] Climbing over parapets under a hail of musket fire, seeing their comrades torn to pieces by artillery, and then enduring exhausting hand-to-hand combat in the streets of the town made Russian soldiers hateful. Potemkin was clearly ashamed of what had taken place under his leadership, for he made no specific mention of the massacre and preferred to concentrate on rewarding the surviving officers and soldiers. 'I have nominated for awards only those about whose merit I am entirely certain,' he wrote to Catherine in March 1791. 'But if, after a close and detailed examination, there are others who are found to be worthy of rewards, then for them too I then dare ask crosses of St George and St Vladimir.'[93]

Finally, the terrible storm of Izmail equipped and updated the Russian army with knowledge of siege warfare at the start of the Revolutionary Wars. The experience was codified in Russian military manuals, most famously by Suvorov in his famous work *The Science of Victory*. Suvorov put that knowledge to use during the bloody siege of Warsaw in 1794 and again during the War of the Second Coalition when he was sent to Italy as the allied commander of Austrian and Russian forces in 1799.

---

92   See for example, Miakinkov, *War and Enlightenment in Russia*, pp.1–4.
93   Quoted in Lopatin, *Potemkin i Suvorov*, p.222.

As Christopher Duffy remarked in wonderment 'By accepted standards of the eighteenth century the allies made remarkably short work of sieges that would reasonably be expected to extend into weeks or into months.'[94] As the Russian Empire entered the military storms of the nineteenth century, its military's approach to sieges, which was rooted in eighteenth-century military culture, was firmly entrenched.

Bibliography

Archives
Russian State Archive of Early Acts (RGADA), fond 1453

Primary Sources
Barskov, Ia. L. (ed.), *Perepiska Moskovskikh masonov XVIII veka* (St Petersburg: Izd. Otd-nie russkago iazyka i slovesnosti Imperatorskoi akademii nauk, 1915)
Chernyshev, Grigorii, 'Pisma vo vremia osady Izmaila 1790 goda. Ot grafa G. I. Chernysheva k kniaziu G. F. Golitsynu', *Russkii arkhiv*, 3 (1871): pp.386–408
Damas, Roger de, *Memoirs of the Comte Roger De Damas (1787-1806)* (London: Chapman and Hall, 1913)
Denisov, Adrian. *Zapiski donskogo atamana* (St Petersburg: VIRD, 2000)
Engelgardt, Lev, *Zapiski L'va Nikolaevicha Engelgardta, 1766-1836* (Moscow: Izdanie Russkago Arkhiva, 1868)
Kutuzov, Mikhail, 'Arkhiv Kniazia M. I Golensheva-Kutuzova-Smolenskogo, 1745-1813', *Russkaia starina* 2, (1870), pp.498–514
Ligne, Charles Joseph De, *The Prince De Ligne: His Memoirs, Letters and Papers* (Boston: Hardy, Pratt & Co, 1899)
Lopatin, Viachislav (ed.), *Ekaterina II i G. A. Potemkin: lichnaia perepiska, 1769-1791* (Moscow: Nauka, 1997)
Meshcheriakov, P., (ed.), *A. V. Suvorov* (Moscow: Voennoye Ministerstvo SSSR, 1949-53)
Mosolov, Sergei, 'Zapiski' *Russkii Arhiv*, 1 (1905), pp.124–173
Polnoe Sobranie Zakonov Rossiiskoi Imperii (St Peterburg: Pechatanov Tip. II Otdieleniia Sobstvennoi Ego Imperatorskago Velichestva Kantseliarii, 1830-1960)
Richelieu, Armand de, 'Journal de mon voyage en Allemange,' *Sbornik Imperatorskago Russkago Istoricheskago Obshchestva*, 54 (1886), pp.111–200

Secondary Sources
Aksan, Virginia H., *Ottoman Wars: An Empire Besieged 1700-1870* (New York: Longman/Pearson, 2007)
Beskrovnyi, L., *Russkaia armiia i flot v vosemnadtsatom veke* (Moscow: Voennoe Izdatelstvo, 1958)

---

94   Christopher Duffy, *Eagles Over the Alps: Suvorov in Italy and Switzerland, 1799* (Chicago: Emperor's Press, 1999), p.118.

Best, Geoffrey, *War and Society in Revolutionary Europe, 1770-1870* (Leicester: Leicester University Press in association with Fontana Paperbacks, 1982)

Brandenburg, N. E., *500-letie russkoi artillerii. 1389-1889* (St Petersburg: Tipografiia 'Artilleriiskogo zhurnala', 1889)

Duffy, Christopher, *Eagles Over the Alps: Suvorov in Italy and Switzerland, 1799* (Chicago: Emperor's Press, 1999)

Duffy, Christopher, *Russia's Military Way to the West: Origins and Nature of Russian Military Power, 1700-1800* (London: Routledge & Kegan Paul, 1981)

Fuller, William C., *Strategy and Power in Russia, 1600-1914* (New York: Free Press, 1992)

Ionin, S. N., *Russkaia artilleriia: ot Moskovskoi Rusi do nashikh dnei* (Moscow: Veche, 2006)

Longworth, Philip, *The Art of Victory: The Life and Achievements of Field Marshal Suvorov, 1729-1800* (New York: Holt, Rinehart and Winston, 1966)

Lopatin, Viachislav, *Potemkin i Suvorov* (Moscow: Nauka, 1992)

Keep, John, *Soldiers of the Tsar: Army and Society in Russia, 1462-1874* (Oxford: Clarendon Press, 1985)

Madariaga, Isabel de, *Russia in the Age of Catherine the Great* (London: Phoenix Press, 2003)

Menning, Bruce W., 'Paul I and Catherine II's Military Legacy, 1762-1801', in Frederick W. Kagan, and Robin Higham (eds.), *The Military History of Tsarist Russia* (New York: Palgrave, 2002), pp.77–105

Menning, Bruce W., 'Train Hard, Fight Easy: the Legacy of A. V. and His 'Art of Victory',' *Air & Space Power Chronicles*, 1986, pp.79–88

Miakinkov, Eugene, *War and Enlightenment in Russia: Military Culture in the Age of Catherine II* (Toronto: University of Toronto Press, 2020)

Mikaberidze, Alexander, *Kutuzov: A Life in War and Peace* (New York: Oxford University Press, 2022)

Nilus, A. A., *Istoriia meterial'noi chasti artillerii* (St Petersburg: Tipografiia P. P. Skoikina, 1904)

Orlov, N. A., *Shturm Izmaila Suvorovym v 1790 godu* (St Petersburg: Sklad izdaniia u V.A. Berezovskago, 1890)

Osipov, K., *Alexander Suvorov* (New York: Hutchinson and Co., 1944)

Petrushevskii, Aleksandr, *Generalisimus kniaz Suvorov* (St Petersburg: Stasulevicha, 1884)

Polevshchikova, Elena, 'Frantsuzkie volontery v Izmaile: neopublikovannaia zapiska grafa Lanzherona', *Deribasovskaia-Reshilevskaia: Literaturno-khudozhestvennyi, istoriko-kraevedcheskii illiustrirovannyi almanakh*, 29 (2007), pp.6–11

Prochko, I. S., *Istoriia Razvitiia Artillerii: S Drevneishikh Vremen i Do Kontsa XIX Veka* (St Petersburg: Poligon 1994)

Sebag, Simon Montefiore, *Prince of Princes: The Life of Potemkin* (New York: Thomas Dunne Books, 2001)

Skritskii, Nikolai, *Georgievskie kavalery pod Andreevskim flagom. Russkie admiraly – kavalery ordena Sviatogo Georgiia I i II stepenei* (Moscow: ZAO Tsentrpoligraf, 2002)

3

Forging a Chain of Security. The Story of the Wellington Barrier, 1815–1830

Beatrice de Graaf

Introduction[1]

Post-war security orders do not rest only on the bayonets of soldiers or the paperwork of treaties, but also need a foundation of the bricks, stones and mortar of defensive lines and fortifications. The security system created by the allied victors of the seventh coalition was no different in this respect. Strangely enough, amongst the countless studies on the Congress of Vienna and the congress system, nothing has ever been published on the pivotal role of the line of fortifications created after 1815, the so-called Wellington Barrière. In this chapter, I will take some clues from the materiality turn, as defined by Bruno Latour as the theoretical turn that emphasises objects, instruments, and other types of 'embodiments' as significant for understanding how organisations and structures function, and what their underpinning ontologies are.[2] Applied here, that means that I will trace how the trajectory of translating hard-won victories, abstract concepts on balance of power and mutual security, into brick and stone played out. We will follow the course of the barges, bringing clay and stone to the Low Countries, where the derelict ruins of the old fortresses of the Vauban era were reconstructed. We will see how emotions, memories, and feelings of revenge and retribution were negotiated, handled and assembled over the course of a few years into a supranational line of defence, the first multinational defence system ever created. The command-and-control centre managing this line of assemblage was the Allied Council in Paris, where the paper-trail leading to the Wellington Barrière started. With these blueprints for security, the allied ministers joined their forces, forged strategic plans and funds together, and intervened directly in the lives of ordinary people. Their international political designs became 'things', a series of objects and stone constructions that disrupted

---

1   This paper is based on, and partly taken from, the research that I did for the monograph Beatrice A. de Graaf, *Fighting terror after Napoleon. How Europe became secure after 1815* (Cambridge: Cambridge University Press, 2022), in particular chapter 8.
2   Bruno Latour, 'Can We Get Our Materialism Back, Please?', *Isis*, 98:1 (2007), pp.138–142.

the lives of citizens. But these same citizens in turn also tried to adjust to and exploit those invading and invasive constructions, and developed social practices that gave different contexts and meanings to the fortresses, sidelining and outwitting the architects of the balance of power to some extent. This chapter investigates how the balance of power *concept* was translated into *physical* stone, it unpacks the Great Powers' tactics of assemblage, and their redirection towards different purposes over the course of construction. In the end, we will ask ourselves what remained of this first European mutually assured defence system: how much materiality was left?

The King and his Fortress

For centuries, fortifications and fortresses were the most visible signs of security in a country. Ramparts and fortified walls marked out the city limits and provided protection to its inhabitants. But they also forced them to bear the costs and consequences of border protection and national defence, such as sieges and artillery bombardment in times of war. One such important line – referred to as a 'barrier' after the seventeenth century – was the Scheldt River, which flows from northern France through Belgium to the North Sea – one of the key lines of the later Wellington Barrier.[3] From early on, castles and ramparts had been built along this river, the town of Oudenaarde being a seminal defence point along this line, since it oversaw the traditional invasion route from France northwards. In 1688, the fortresses here came into the hands of Louis XIV, who charged his brilliant military strategist and designer of fortifications, Vauban, to turn Oudenaarde into a bastioned border town. A number of dramatic sieges later, the Barrier Treaty of 1715 gave the Dutch Republic the right to establish garrisons in the barrier fortresses in order to prevent France from returning to 'its politics of dynastic imperialism', thus transforming Oudenaarde's ramparts into a visible sign of the European system of a balance expressed in stone and border fortifications.[4] That system lasted less than half a century; after 1748, Vauban's bastioned fortifications slowly fell into disrepair. The barrier cities were allowed to sell the land to local inhabitants on the condition that they would dismantle the fortresses. By 1803, there was little left of the other barrier fortifications.[5]

Old times returned in 1814. During the last months of the Napoleonic Wars, 180,000 soldiers streamed through the town that almost collapsed under the burden

---

[3] See Robert Gils, *De versterkingen van de Wellingtonbarrière in Oost-Vlaanderen* (Ghent: Provincie Oost-Vlaanderen, 2005); Pol Borremans, *Het Kezelfort van de vesting Oudenaarde* (Erpe: Krijger, 2006). One often thinks of Antwerp when talking about the Scheldt, and the Eastern and Western Scheldt. But the Scheldt estuary has many tributaries, like the Lys (Leie), which flows into the Scheldt in Ghent, or the Dender, which joins the Scheldt in Dendermonde, flowing eastwards from there. These were important defensive positions.

[4] Henri and Jacques Pirenne, *Geschiedenis van Europa*. Vol 2: Ontstaan en groei der westerse beschaving (Brussel: La Renaissance du Livre, 1966), p.182.

[5] Borremans, *Het Kezelfort van de vesting Oudenaarde*, pp.7–17. For Ypres, see: Rik Opsommer, *Ieper en de Frans-Belgische grens (17e–18e eeuw): 300 jaar vredesverdragen van Utrecht en Rastatt* (Ypres: City Archives, 2013).

of occupation, billeting, and the need for fodder and other supplies. But Oudenaarde was to be restored to its former splendour. At the end of March 1814, the Allies decided that the old barrier fortifications needed to be reconstructed and rebuilt. The Kingdom of the Netherlands, which had yet to be formally established, was to line its border with France with a wall of fortified cities (fortresses) and Oudenaarde should become part of the new fortification plan, which was later named in honour of the allied commander who 'won' Waterloo, the Duke of Wellington – the Wellington Barrier.[6] Indeed, the landowners on the Kezelberg had their properties expropriated, the fortification work was contracted, and stones of Boom clay – a very hard material from the Scheldt found at brickyards in the area – were hoisted up the mountain with concerted effort.[7] On 27 June 1823, the completion of the fortress, the Kezelfort, the most lavish of all fortresses involved, was to be celebrated and the fortress itself inaugurated by the King himself, Willem I. All the inhabitants were standing on the slopes of the Kezelberg; they had set up beautifully decorated tents, with wine to honour the special occasion. The King, together with *Colonel* Hennequin from Ghent, the head of the third Directorate of Fortifications, studied the building plans, took a look at the passageways and fortified chambers – casemates – that had been dug into the hillside and secured with skilful geometric masonry. But then things went awry. De Rantere, the chronicler of Oudenaarde, writes: 'A few minutes after the king had emerged from one casemate, a good amount of dirt caved in on the frontside of that very casemate, much to the great surprise of the spectators.' The small landslide caused a good deal of hilarity among the onlookers, but Willem was not amused. 'The king was offered a skilfully carved stone with his name and the date on it to install in a prepared opening in the tower's wall. But he chose not to do so'.[8] An investigation into the mishap was initiated, leading to the arrest and suicide of the said *Colonel* Hennequin and the dismissal of some prominent architects. More on that later. To this day, during carnival, the citizens of Oudenaarde search for that royal stone, which disappeared in the commotion of the landslide and has never surfaced again.

What happened here? The materiality of the fortress did not uphold the reputation of King Willem I and the original intentions of the Big Four, the Allied Council in Paris. It was the sand and stone delivered by the indigenous Belgian contractors, who had 'played' the allied and royal clients and had redirected the building contracts to line their own pockets. Instead of producing expensive bricks, they had filled the inner fortress walls with sand, keeping the surplus budget for themselves. The Wellington Barrier was intended as a visible, material and concrete symbol for the new balance of power, the product of the Allies' cooperation and their conquest of France. As Hobbes had it, 'The reputation of power is power'. Indeed, 'Power

---

6   Borremans, *Kezelfort*, pp.17–22.
7   City of Oudenaarde, Modern Archive, Deliberations of the city council, 3 Janury 1820 – 24 September 1825, SAO. 'Onthaal van den Koning', meeting of the government's council, 21 June and 1 July 1823.
8   M. de Smet (ed.), *Het dagboek van Bartolomeus de Rantere. Beschrijving van al het merkweerdigste dat er voorgevallen is in de stad en casselrije van Audenaerde zedert het jaer zeventhien hondert zevenentachentig tot het jaer achthien hondert vijfentwintig* (Oudenaarde: Sanderus, 1973), pp.240-242.

and authority cannot endure without collective imaginations and ascriptions'.[9] The allied power had to be visible, even after the large allied occupation force left France. But the only visibility here was the risibility of a King startled by the collapse of his first fortress – an omen for future defeats in Belgium. But first of all, what were the ideas and designs behind the construction of this vast defence project?

From Revenge to Occupation

After Napoleon's second defeat, the allied victors did not go home but decided to stay in Paris, working together on the negotiation of a second Treaty of Paris and remaining until all treaty stipulations were met. The predominant danger that needed to be deterred and averted was the rise of another wave of terror – defined in 1815 as a two-headed monster of Napoleonic despotism and hegemonic aggression on the one hand and revolutionary upheaval on the other. As the vast body of literature on the Concert of Europe had it, Vienna laid the foundations of a newfound balance of power, an equilibrium between Britain and Russia, to which the other larger and smaller powers could subscribe. New insights, after Paul Schroeder, have moreover convincingly argued that norms and institutions produced in Vienna paved the way for the creation of a new security culture of deliberation and negotiation, rather than remaining stuck in the floating and unstable alliances and cabinet wars of the eighteenth century. Elsewhere, I have argued that this literature has not made clear how this transition from unstable wartime alliances gave way to a concert of deliberation and consultation that lasted throughout the nineteenth century. Pointing to Vienna alone misses the glaring fact that the real conferences and treaties were only concluded after Napoleon's return, and his second defeat. Rather than going home after Waterloo, this time the powers decided that the paperwork of treaties was not enough to prevent the monster of terror from raising its ugly head again. They needed to devise another scheme to deal with France, as a country and as a power, and with the political ideology of Bonapartism and the revolution. For this reason, they opened the Allied Council, the Conference of Allied Ministers in Paris, which convened from 12 July onwards, on a daily basis, from 1816 twice weekly, and lasted in this form until December 1818. After that, it also continued, but on a level lower, an ambassadorial level. The literature on peace and security in Europe needs to be revisited to introduce this seminal transitory period, which laid the groundwork for the system of conferences and multinational security operations. The first conference being dedicated to the allied occupation of France, the fight against terror, and the construction of a new and material system of defence to prepare against war in peacetime.

The first obstacle to such a multinational system of defence was the thirst for revenge on the part of the Prussians. The Prussian generals, Blücher and Gneisenau, were already frustrated that Wellington prevented them from executing Napoleon.

---

9   Barbara Stollberg-Rilinger, 'State and Political History in a Culturist Perspective', in A. Flüchter and S. Richter (eds.) *Structures on the Move: Technologies of governance in transcultural encounter* (Berlin: Springer, 2012), p.47.

Hardenberg, less of a military bully and more the bureaucrat, was, however, equally demanding: he submitted a Prussian claim for 1,200 million francs in damages from France (adding war contributions imposed by Napoleon, requisitions by the French, and reparation for the damage suffered through the last campaign of 1815).[10] That amount was excessively high – it far exceeded the gross domestic product of France – and did not even include the cost of maintaining the occupational forces.[11] Moreover, on top of these payments, more territories needed to be carved out of France to satisfy the Prussian longing for security from future French invasions. The solution was obvious, according to the strategically well-versed Knesebeck: Alsace should be divided between Prussia and Austria; Prussia should be given the French fortresses along the Mosel and the Saar; Savoy should be governed by the king of Sardinia; and Switzerland should get part of Franche-Comté. As noted previously, the Prussians were in favour of cutting away large swathes of French territory. For clarity's sake, Knesebeck had drawn a blue line on a map of France to indicate the width of those swathes. A stretch on the north and east side that included Lille, Mézières, Montmédy, Metz, Strasbourg, Colmar and further to the south Châtillon and Barraux would be cut off from France. This would create a 'good line of defence' and the French fortresses would be in the hands of neighbouring countries. When coupled with occupation forces in the heart of France for the first years, peace would certainly be assured.[12] 'Today, we can still do this', opined Knesebeck (who, here too, expressed Blücher's firm conviction). 'The hand of providence has demonstrably granted us this opportunity. If we let this chance pass, rivers of blood will flow later to attain the same goal, and the groaning of the hapless ones will call us to account!'[13]

Remarkably all other allies were dead set against such demands of conquest. The other ministers felt that the Allied Council should not fan feelings of revenge, but that they needed to 'curb the subversive principles that undermine the social order and on which Bonaparte had based his usurpation'.[14] In other words, for Prussian commanders and ministers to demand major territorial revisions and exorbitant indemnities would only feed the spectre of terror. The extent of the reparations had to be such that it did not provoke an unnecessary amount of instability in France.[15]

Then, Metternich, supported by the British, made an interesting move. He suggested limiting financial reparations to a more moderate figure and, in return, introduced the subject of fortifications into the discussion. The potential for French insurgence should not only be kept in line with financial 'securities', but could also be addressed by removing the French line of defence along its borders.[16] The Allies had to 'help' France transition from an offensive to a defensive security policy. Crucial

---

10 Geheimes Staatsarchiv Preussischer Kulturbesitz, Berlin (GStA): PK, III. HA. I. Nr. 1461. Sum of the Balance, July 1815. Probably compiled by Hardenberg.
11 GstA: III HA.I. Nr. 1461, pp.65–66. Hardenberg, *Mémoire*, 22 July 1815, p.27.
12 GStA: III HA.I. Nr. 1461, pp.65–66. Hardenberg, *Mémoire*, 22 July 1815; GStA: III HA.I. Nr. 1461, p.68. See also pp.69–73, Knesebeck, Memorandum, 4 August 1815.
13 GStA: III HA.I. Nr. 1461, pp.69–73.
14 GStA: III HA.I. Nr. 1461, p.75.
15 GStA: III HA.I. Nr. 1461, p.75.
16 GStA PK, III. HA I. no. 1461. Metternich, 'Memorandum', 6 August 1815. Sent to Hardenberg, p.75.

to this transition was the line of forts along France's northern and eastern border, which, positioned as they were so close to the neighbouring countries' territorial borders, could never serve as an exclusively defensive presence. Given the balance of power framework, it was logical, according to Metternich, that this French system of fortifications should be modified – especially because all the forts in the Low Countries and Germany were destroyed or in a state of disrepair. It would take far too long and be too expensive to wait until those countries could restore their fortifications. So, in anticipation of the reconstruction of the strongholds in these neighbouring countries, France could begin to help pay for new fortresses, or possibly also dismantle their own fortifications or transfer them to the Allies.[17]

Russia and Britain agreed. The decision was made, and formalised in the Treaty of Paris of November 1815. France was to be occupied, not exceeding a period of five to seven years, and French reparations should be enough, but not exceed, the cost of a new line of defence for the protection of neighbouring countries and the Allies' occupation of French fortresses until that new line was complete.[18] This was justified with the claim that 'within just limits the Allies are entitled to the fruits of conquest, and therefore to such permanent acquisitions as they might deem necessary for their own security'.[19] This combination of restraint in financial demands and territorial claims was not self-evident. Friedrich von Gentz, Metternich's assistant, knew that the majority of the British population was in favour of imposing hefty reparations, both in land and in capital. Public outrage in the United Kingdom was tamed only because of Wellington's shining reputation on the home front.[20]

In short, the allied council had to translate the principle of balance of power into a system guaranteeing collective security amongst the powers, on the one hand directed against France, but on the other, without pushing France entirely overboard. The focus was not revenge, dismemberment or bleeding France to death financially, but designing a balanced system of payments. The brilliance of the allied plans lay in the fact that French reparation payments would be funnelled into the reinforcement and reconstruction of the fortress barrier intended to create a new defence line against France. This line of forts, fortresses and fortified cities would serve the European balance of power, as Pitt had anticipated back in 1805. The future barrier would fulfil a threefold function: act as a deterrent against renewed French aggression, as a buffer between Prussia and the North Sea and, demonstrating the special relationship between the Netherlands and Great Britain, serve as a sign of Britain's advanced sphere of influence on the continent.[21] Interestingly enough, the

---

17   GStA PK, III. HA I. no. 1461, pp.77, 80.
18   Enclosure ('an official memorandum in consequence of the Russian paper'), included in Liverpool's letter to Castlereagh, 3 August 1815, in: A. Wellesley (ed.), *Supplementary Despatches and Memoranda of Field Marshal Arthur Wellesley, 1st Duke of Wellington* (hereafter: WSD) (London: John Murray, 1871), vol.11, pp.86–89.
19   WSD, vol.11, p.87.
20   Friedrich von Gentz to Karadja, 5 September 1815, in F. von Gentz, *Dépêches inédites du chevalier Gentz aux Hospodars de Valachie: pour servir à l'historie de politique Euroopéenne* (Paris: E. Plon, 1876), vol.1, pp.172–175.
21   See Niek van Sas, *Onze natuurlijkste bondgenoot. Nederland, Engeland en Europa, 1713–1831* (Groningen: Wolters-Noordhoff, 1985).

defence design was thus very much informed by the balance of power considerations of 1814/1815. It was as much a device to keep the western allies together (and thereby indirectly, a deterrence against Russia, which had moved very far into the West indeed with its troops occupying France as the negotiations took place), as it was a way to avoid entering into excessive obligations on the continent, bringing the British troops back home, and having the fortresses with Dutch and German troops instead as a deterrent to France.

Boulevard de l'Europe

The first visible signs of the Allies' security order, apart from the occupation army, were to be the new forts outside France. But before these new strongholds could be reinforced or rebuilt, the operation's first step was the evacuation and dismantling of the French fortresses. This was not an easy task, nor did it totally succeed. The importance of controlling these fortresses in the French border region had become evident in July 1815. Despite Napoleon's capitulation, many fortified towns in the north and east of France refused to surrender to the Allies. The agreement was that French troops were to remove themselves to south of the Loire, where Napoleon's *Grande Armée* would be disbanded and then be reorganised again under the royal banner. But pockets of resistance remained north of the Loire, particularly in places where the French troops could count on the 'patriotism' of inhabitants along the border.[22] French garrison troops from the fortresses of Mézières and Longwy attacked the allied Prussian troops time and again.[23] In addition, the garrisons in the fortified towns would not let the allied troops pass through those towns, which meant that supply and transit routes for the (until January 1816) approximately one million allied troops were obstructed or blocked. The military committee, chaired by Wellington,[24], therefore, quickly issued clearer orders: the French troops had to leave the occupied part of France, and do so immediately. The fortresses had to be vacated, pending further negotiations about France's borders and the eradication of the forts or their allocation to either France or neighbouring countries.[25] It took many more months. Cities such as Vitry, Metz, Thionville and Verdun complied with the summons in July and August. At the end of July, skirmishes were still common around Auxonne and Mézières. The population and troops in and around the citadel of Briançon kept up their resistance through the beginning of 1816. When they finally left these towns, the French troops tried to take whatever they could with them, including cattle and victuals. They also tried to destroy the ammunition in the depots or to move it elsewhere; anything to keep it from falling into the hands of the Allies.

---

22 See Volker Wacker, *Die alliierte Besetzung Frankreichs in den Jahren 1814 bis 1818* (Hamburg: Kovac, 2001), p.143.
23 GStA III Nr. 1465. Report Frederik of Orange and Wellington. Discussion led by Wellington in the Allied Council, 18 August 1815.
24 The other committee members were the allied generals Schwarzenberg, Gneisenau, Wolkonsky, Radetzky and De Wrede.
25 GStA III Nr. 1464. Protocol 16, 17 July 1815.

For both sides, the Allies and the French, the fortresses were a visible and volatile symbol, a matter of prestige. The French saw the fortified ramparts along their border as an expression of sovereignty, as they vehemently expressed by the mouths of Talleyrand and later Richelieu during the sessions of the Allied Council. For the Allies, the surrender of the forts was proof of France's submission to their authority and the first step to erecting their own, anti-French, European barrier of security. That is why the allied ministers, during their negotiations on the second Treaty of Paris in September 1815, were in complete agreement on this point. Many of the fortifications on France's side of its northern and eastern border had to be demolished, unless they were transferred to neighbouring countries. The military line of occupied France would run along the forts of Condé, Valenciennes, Bouchain, Cambrai, Lequesnoy Maubeuge, Landrecy, Asesnes, Rocroy, Givet, Mézières, Sedan, Montmedy, Thionville, Longy, Bitsch and Fort Louis. The Peace Treaty required the return to the territorial borders as they were in 1790, which meant that the Landau, Saarlouis, Philippeville, Marienbourge and Versoix fortresses would be transferred to the Allies (that is, to the Netherlands and Prussia). The appropriated areas came nowhere close to the wide swath that Prussia wanted to see excised from France.[26] Sardinia and Switzerland each also received some territory.

With this evacuation ongoing, the next step was to be the setup of a managing board for the construction works, a supranational project, the European bulwark, ranging from the North Sea, via Namur and Mainz, to Switzerland and the Mediterranean. The Treaty of Paris also saw to the financial footing of the project. A sum of 187.5 million francs in reparations would be reserved for the construction of fortifications on the Dutch and German side of the border. Because Saarlouis eventually went to Prussia (as previously noted), 50 million was deducted from that fortress fund, leaving 137.5 million francs.[27] Of that amount, 60 million was earmarked for fortifications in the Netherlands. The other 77.5 million francs were distributed among fortifications in the Bas-Rhin, Haut-Rhin and Piedmont regions. Spain, too, would in due course, after signing the Treaty of Vienna, receive a few million francs for work on fortifications along its border.[28]

The significance that the allies attached to the matter of fortifications was moreover translated in the appointment of Wellington as commander-in-chief of the allied army of occupation, chair of the military subcommittee to the Allied Council, and, thereby, as manager-in-chief of the fortification project – meaning that he would also manage the fortress fund. In practice, that meant that requests would be discussed by the Allied Council, but the final approval and allocation of funds would run via Wellington. Practically speaking, this meant that the countries bordering France, and the Netherlands and Prussia in particular, could request

---

26 GStA PK, III. HA I. no. 1461, pp.65–66. See map in the Knesebeck, Memorandum, 4 August 1815.
27 GStA III Nr. 1469. Decision 19 September 1815. GStA III Nr. 1465. See also: protocol 2 October 1815.
28 GStA III Nr. 1469. Protocol 7 and 8 October 1815. Territorial additions to the Netherlands, Switzerland and Sardinia were adopted on 6 November and added to the Treaty of Paris. Protocol 6 November 1815.

funds to finance the construction of their fortifications. The Council appointed special commissioners who would issue the vouchers for paying for the fortifications and see to it that these funds were actually used for the intended purpose.[29] According to the protocols of the Allied Council, the Dutch requests were always granted. In May 1816, however, Sardinia experienced a rude awakening because it had asked for too much money. The King of Sardinia had been promised 10 million, but he now threatened to exceed that amount. The claims from the small German states did comply with what had been agreed upon, and were also honoured. The funds were distributed discriminately.[30] In March 1816, Prussia tried to get the cash register and the funds for the fortresses moved to Frankfurt, because, at that time, the largest building projects were taking place there. Austria, too, thought that the funds belonged to the country in which the building projects were to take place. But the other ministers foiled that argument, pointing out that Spain, Sardinia and the Netherlands had projects as well. The fund was a pool of resources held in common by the 'princes of the coalition'; its management remained in Paris.[31]

In other words, building a visible and tangible European bulwark, a system of defence was in the hands of the Prince of Waterloo, which only helped to increase his power and influence even more. His overriding influence also expressed itself in the Allies' strategic planning and in the symbolism of the 'bulwark of Europe' that would be set up along a line from the North Sea to Mainz – a presence that in Belgium was soon christened the 'Wellington Barrier'.[32] It was fitting that an Englishman oversaw the barrier, since the Pitt Memorandum of 1805 had revived and underlined the Peace of Utrecht's barrier treaty of 1713 again. With this memorandum, the role of the Netherlands was designated to function as a buffer state, a conduit and a unified defence line against France – as a 'Boulevard de l'Europe', a bulwark of Europe.[33] A defensive line running from the Lower Countries, the Rhine States of the German Federation, Switzerland, Piedmont-Sardinia and Spain was to be the crucial military underpinning for the restoration of the European balance of power. These medium-sized countries were themselves too small to pose a threat to France, and too large to be easily trampled underfoot in a quick strike. But taken together, they formed a belt around France. Troops garrisoned here could also be quickly reinforced by Britain, Prussia and Austria, and served as a buffer for the great powers against the possibility of renewed French aggression. As an independent state just across the English Channel, the Netherlands clearly fell within Britain's sphere of influence, was a bridgehead for British troops, a buffer between Prussia and the North Sea, and a stronghold against France. Although the Netherlands would never be able to completely withstand a strong French offensive,

---

29 The National Archives, Kew (TNA): Foreign Office 146/15. Protocol 24 November 1816.
30 See, for example, protocol 26 May 1816 in which the additional claim from Sardinia is discussed and rejected.
31 TNA: FO 146/6. Protocol 10 March, 2 June 1816.
32 See TNA: FO 92/31, pp.77–80. Wellington, memorandum, 22 July 1816; FO 92/31, pp.93–97. Fagel to Wellington, 18 October 1816.
33 For the early history of the barrier, see: R. Geikie and I. Montgomery, *The Dutch Barrier, 1705–1719* (London: Cambridge University Press, 1930).

the hope was that it could resist an attack long enough to give British or Prussian troops time to arrive.³⁴

The Uses of Fortresses

A brief aside should be addressed to the question of how useful the construction of fortifications was in light of the new military reforms and methods of the Napoleonic era. It had been demonstrated during the recent wars that all the new roads that had been built in the eighteenth century could, in times of war, never be closed off effectively with barriers or fortresses. Armies were also much larger, such that one could easily put a fortress under siege with some of your troops, while the rest of the main army could make its way around it. Warfare that focused on fortresses, with sieges or battles in the open field, had become more challenging. Improved artillery trains, the greater agility of armies and more convenient ways of supplying provisions allowed army corps to manoeuvre with greater ease and speed. Napoleon had already shown how quickly large armies could move, simply leaving fortresses to stand where they were, and to force the enemy to engage them on a battlefield of his choosing.³⁵

That did not mean that the role of fortresses was altogether passé, but it was easier simply to ignore them. In other words, the eighteenth-century tactics of sieges and battles belonged to the past rather than to the future of warfare. Wellington was himself the first to admit that, from the military-technical point of view, building fortresses was somewhat outdated. On 22 September 1814, in a memorandum to Earl Bathurst, Secretary of State for War, Wellington did not shy from sharing his hesitations about investing in new fortresses: 'The operations of the revolutionary war have tended in some degree to put strong places out of fashion'. The recent campaigns against Napoleon had also shown 'that strong places are but little useful, and at all events are not worth the expense which they cost'.³⁶

Nevertheless, Wellington himself knew from his campaigns in the Peninsula how important fortresses could be in absorbing the attention of a large number of troops. Moreover, the fortresses needed to be built, not solely from a military point of view,

---

34 Van Sas, *Onze Natuurlijkste Bondgenoot*, pp.41, 55. See also Gils, *De versterkingen van de Wellingtonbarrière*, pp.8–9. And, of course, G.J. Renier, *Great Britain and the Establishment of the Kingdom of the Netherlands, 1813–1815: A Study in British Foreign Policy* (London: Allen & Unwin, 1930). See also C. Nelson, 'The Duke of Wellington and the Barrier Fortresses after Waterloo', *Journal of the Society for Army Historical Research*, vol.42, no.169 (March 1964), pp.36–43.
35 See Wilfried Uitterhoeve, *Cornelis Kraijenhoff 1758–1840: Een loopbaan onder vijf regeervormen* (Nijmegen: Vantilt, 2009), pp.289–290; J.E. Kaufmann and H.W. Kaufmann, *Fort and Fortifications of Europe, 1815–1945: The Central States* (Barnsley: Pen and Sword, 2014), pp.5–6. See also idem, vol 2. *The Neutral States* (Barnsley: Pen and Sword, 2014), pp.81–82.
36 Wellington to Bathurst, 'Memorandum on the Frontier of the Defence of the Netherlands', Paris, 22 September 1814, in J. Gurwood (ed.), *The Dispatches of Field Marshal the Duke of Wellington, K. G. during his Various Campaigns in India, Denmark, Portugal, Spain, the Low Countries and France* (hereafter: WD) (London: John Murray, 1838), vol.12, p.126.

but also from an internal and external security perspective. Internally, because they could serve as a base to put down an insurgency against the occupying forces. Additionally, during the occupation, with an eye to logistics and maintaining open supply lines for the occupying army. Control of the access roads leading into and out of France, and hence of the fortifications along the border, was indeed crucial for the Allies. Externally, because there was no other way to defend the countries to the north and east of France. The small effort required to restore the existing fortifications would, in any case, slow the French down significantly should they decide to invade the country again. Improving those sites would increase the likelihood of a protracted and more gruelling confrontation for the enemy, and was hence still a useful and even necessary investment.[37] From an operational point of view, these investments were furthermore worth the effort when these fortresses were constructed in places that were important for the effective use of the field army. When situated more spread-out in the back country, they provided a safe operating base for the troops in the field, as well as a safe haven to which they could withdraw should they have to recuperate from defeat elsewhere (for example, the ring of forts – the National Redoubt – built around Antwerp).[38]

But perhaps most importantly, the fortresses were the first material operation of a common allied security policy in peacetime. It was, along with Gruner's security service, Metternich's passports and the occupation army itself, the most tangible and, so it was thought, sustainable way of binding the Quadruple Alliance together. It was a pledge in stone of their reciprocal security obligations and a visible manifestation of the allied concert. On 21 November 1815, one day after the signing of the Treaty of Paris, the ministers agreed that they would jointly maintain this 'essentially European system'. The great powers would jointly supervise the 'defensive line of countries bordering France'.[39]

The significance of taking on the construction of such an immense line of defence along the French border as a joint project of the Allied Council lay in the fact that it decreased the risk of the Alliance disintegrating, of individual countries making separate treaties with France or of carrying out any self-willed aggression. The fortress fund and Wellington's supervision were an effective guarantee of the Allies' solidarity when it came to security policies. Their pledge to finance and build these fortifications bound them together.

Lapdog and Net-Recipient

Secondary powers were given a major role in implementing the fortifications project. That was especially true for the Netherlands. Back in Vienna, Metternich had informed King Willem I's envoy, Baron von Gagern, that he saw 'Holland' as the 'lapdog of the great powers'. Though meant to humiliate, that pat on the head worked to the Netherlands' advantage: independence, the establishment of a

---

37 WD, vol.12, pp.127–129.
38 Gils, *De Versterkingen van de Wellingtonbarrière*, p.9.
39 GStA HA III Nr. 1469. Agreements 21 November 1815.

monarchy and territorial expansion. The Netherlands also became the largest net recipient of money from the fortress fund, even though this handout would come to hang like a millstone around the neck of Willem I.

Wellington had already laid out the exact plan for such a 'pledge in stone' of allied guarantees for the European collective security, by means of a ring of fortresses from the North Sea running through the southern frontier of the Netherlands to Germany and the south. In 1814 he had presented his ideas to the Prince of Orange in a 'Memorandum on the defence of the Frontier of the Netherlands', outlining how the unification of the northern and the southern Netherlands (current day Netherlands and Belgium) was necessary to take care for 'additional security to its frontier, by placing in the hands of the government of the Dutch Provinces, those countries which were always deemed essential to their defence and from the whole form a State on the Northern frontier of France which by its resources, its military strength & situation should be a Bulwark to Europe on that side'. This frontier should moreover serve as a conduit between Britain, the Netherlands and Germany: 'The secure Communications of them with England, & the north of Germany are essential objects in any system of defence to be adopted & above all that with Breda & Bergen op Zoom and with the Dutch places on the lower Meuse, and lower Rhine'.[40] This same plan gained even more strength after the 100 days period and was put on the table again in 1815 – this time with a budget to support it.

The second Treaty of Paris, which included special provisions for a fortress fund, helped to resolve the pecuniary problems relating to the construction of the frontier. As noted previously, 60 million francs went to the Netherlands, and Great Britain contributed an additional two million pounds sterling as compensation for the Cape Colony and possessions on the north coast of South America,[41] which the Dutch had officially ceded to Britain in the Anglo-Dutch Treaty of 1814. During the Napoleonic occupation of the Netherlands, the British had taken possession of these colonies. The Netherlands agreed to invest two million pounds of its own in the barrier, while London indicated that its share should not exceed three million. Wellington managed the funds, and all the expenditures and budgets were to be controlled and approved by him. The construction would be carried out by Dutch and British engineers together, with Prussia advising when it came to Luxembourg (and funded in part by the French).[42] The fortification projects were, at the time then, clearly not a manifestation of Dutch independence, but more a confirmation of its role as lapdog – or better put, as a watchdog for the Allies. Willem was thus forced to go along with Wellington's plans and to contribute to the barrier with his own funds as well. The

---

40 Royal House Archives, The Hague, The Netherlands, A 40, Willem Frederik George Lodewijk, Koning der Nederlanden 1792–1849 XIII. Military Affairs, Inv.nr. 14. Wellington, 'Memorandum on the defence of the Frontier of the Netherlands', handed to the Sovereign Prince of the Netherlands. See also H.D. Jones (ed.), *Reports relating to the re-establishment of the fortresses in the Netherlands from 1814 to 1830* (1861), pp.3–9. See also the MA-thesis by Frederik Frank Sterkenburgh, *Van bufferstaat tot neutraliteit, of: De militaire carrière van prins Frederik der NEderlanden, 1813–1840* (Amsterdam: University of Amsterdam, 2012), pp.21–22. With thanks to Frederik Frank Sterkenburgh who sent me his MA-thesis in July 2022.
41 Demerara, Essequibo and Berbice.
42 Gils, *De Versterkingen van de Wellingtonbarrière*, pp.13–16.

entire endeavour would end up costing 70 million guilders (1 guilder was 9.6 grams of silver or 0.6 grams of gold around 1817).[43] As such, more than half of the Dutch defence budget went into the project.

Contesting Wellington

Where the major allies all tacitly agreed, secondary powers could, however, put up some protest and were able to redirect the original allied plans.

In the Netherlands, Inspector General Cornelis Krayenhoff was appointed by King Willem I to conduct the Dutch part of the project. Krayenhoff was a kind of jack-of-all-trades and celebrated hydraulic engineer among his contemporaries. With degrees in law, philosophy and medicine, he had made a name for himself during his Batavian and French period as a physicist. He was an authority on electricity and lightning, had reorganised the Department of Water Management and begun his life's work: mapping the Netherlands in detail using a system of triangulation. His sons also served under Napoleon and worked as far away as Moscow. However, after Napoleon's defeat at Leipzig in the Battle of the Nations, Krayenhoff chose the side of the Allies. Willem appointed him inspector general of all fortifications on 12 March 1814 and bestowed on him the title of baron.

Due to his expertise, he quickly found admiration and trust in British circles. In October 1815, he presented his first sketches to Willem I and Wellington.[44] According to the British ambassador Clancarty, Krayenhoff, who was enjoying this new, prominent position and his proximity to the heroes of Waterloo, was feeling like 'a second Vauban'. On 19 and 20 December, Wellington and Krayenhoff devoted two days to discussions related to the construction of the first fortifications.[45]

Wellington and Krayenhoff agreed straightaway on the first line of defence. It should run via Antwerp, Ostend, Ypres, Menin, Tournai, Ath, Charleroi and Namur to Maastricht. According to Krayenhoff, the costs would amount to 42 million guilders, as he put it: 'just close your eyes and pay the price.'[46] Binche, Marienbourge, Philippeville, Dinant, Bouillon and Arlon were added to this line of defence. Arlon and Binche were never fortified in the end, and Wellington's wish to reinforce Courtrai also did not happen, owing to the overly high cost. Regarding the second line, however, an extended difference of opinion arose between the British and the Dutch that could be traced back to diverging strategic intentions for the bulwark of Europe. Wellington wanted a second line of defence close to the west coast, from Dendermonde via Ghent, Ostend, Oudenaarde and – with an eye to defending Brussels – via Halle to Mont-Saint-Jean, near Waterloo. But Krayenhoff and the Dutch were afraid that, should defeat (in a hypothetical war) appear imminent, the British could too easily leave the area quickly by embarking at Ostend. They

---

43 Uitterhoeve, *Cornelis Kraijenhoff*, pp.292–293.
44 TNA: WO 55/1553/9. Inspector General of Fortification, State of the Fortresses in Flanders, November 1815. Prepared by door Captain Wedekind, on behalf of Wellington.
45 Uitterhoeve, *Cornelis Kraijenhoff*, p.296.
46 Uitterhoeve, *Cornelis Kraijenhoff*, p.296.

therefore proposed a different second line: via Oudenaarde, Dendermonde, Antwerp to Maastricht, with a road and a fortress near Hasselt. Krayenhoff also found that Brussels was insufficiently covered by fortifications at Halle and Waterloo.

On paper, Wellington prevailed. Ghent, Oudenaarde and Dendermonde would be reinforced, but Krayenhoff managed to postpone the planned construction of fortifications at Halle and Mont-Saint-Jean. In the end, nothing came of reinforced fortifications around the fields of Waterloo. To appease Wellington's wishes somewhat, but to spare the Dutch treasury, only a few field fortifications – actually earthworks – were introduced. The defence of the Rupel-Dyle-Demer valleys, between Maastricht and Antwerp, were also put on the back burner, and the Fortifications Service did not even start on the fort at Oosterlo. After the occupation troops left France, Wellington returned to London, where he joined the Liverpool cabinet as Master General of the Ordnance – one of the highest military positions – in December 1818. As a result, he could not involve himself nearly as much with the fortifications. But he never completely lost interest. In 1821, he again conducted an inspection tour, and continued to correspond with Krayenhoff for a number of years.[47]

And so it happened that between 1815 and 1824 a defensive barrier took shape, consisting of 21 fortresses, including the one manned by Germans in Luxembourg. The fortifications were situated along two lines parallel to but also at right angles to the French border. The right-angled line was British, the parallel line, Dutch. From a Dutch perspective, the first line was along the French border and ran from Nieuwpoort on the coast via Ypres, Menin, Tournai, Oudenaarde, Ath, Mons, Charleroi, Namur to Dinant (with Philippeville and Marienbourge as forward posts). In this way, all the major access roads to the Netherlands could be blocked (with Bouillon as an outpost of the Luxembourg fortress). The second line, behind it, ran along the Ghent-Ostend canal and further along the Scheldt and the Rupel-Dyle-Demer valleys in the direction of Maastricht. The citadels of Ghent and Dendermonde were part of this line. A third line, even further north, was located along the southern edge of the northern Dutch territory, running from Zeeschelde via Bergen op Zoom, Breda, 's-Hertogenbosch to Venlo. From a British perspective, however, the fortifications were lying along two operational lines that needed to be defended – the Meuse and Scheldt rivers, which were at right angles to the border with France. Troops and goods could then easily be supplied via these rivers, since there were no railways yet.[48]

Subverting Wellington, and Krayenhoff

The fortifications were meant to defend the new European order. This grounding for Europe's balance of power, however, was built without taking into account the wishes of the inhabitants of the border region. Little or no consideration was given to their possessions, fears and worries. The allied security culture led to more uncertainty and disruption for the citizens in those regions. The last part of this chapter

---

47 Gils, *De Versterkingen van de Wellingtonbarrière*, pp.14–18; Uitterhoeve, *Kraijenhoff*, pp.296–314.
48 Gils, *De Versterkingen van de Wellingtonbarrière*, pp.18–19.

on the chain of security is devoted to the ways and means they tried to subvert the building programme and avoid its inconveniences.

The inhabitants of Oudenaarde had already experienced the inconvenience of military occupation in 1813–1815. They had to hand over carts, horses and supplies, and endure the billeting of French, British, Dutch and German armies. Wellington himself had visited the place on 20 April 1815 to inspect their fortress, and in early June, just prior to the Battle of Waterloo, had explained to the people that he was commander-in-chief of the allied troops. In his statement, he called on the local administrators of Oudenaarde and its residents to report or detain deserting soldiers; an order that was softened with the promise of a reward of 23 Belgian francs per deserter.[49]

When the construction of the Wellington Barrier in the southern Netherlands actually began in late 1815, the local population experienced the move as a great disruption and northern invasion into their region. Work on the fortifications was carried out by local contractors, but they were at the beck and call of the Fortifications Service, and under the supervision of Inspector General Krayenhoff. His department had six directorates, including Ghent, Antwerp and Namur, each with a colonel director at its head. All sorts of officers, engineers, overseers, but also military guards – field wardens, one might say – caretakers and other personnel were stationed at each of the fortresses. Most of them came from the northern, Dutch part of the country. Among the 128 Dutch officers/engineers in 1830, only six were Belgian. The few (French-speaking) Belgian soldiers and engineers who were allowed to join in complained that they first had to learn Dutch.[50]

But there was more to it than that. In early 1816, the provincial governors began announcing expropriations on a large scale. Prior to Napoleon, cities saw to the maintenance of their own fortifications. The monarch, of course, could involve himself in their form and design, but in peacetime, the city magistrate could decide for himself what he did with them. He could break them down, rent them, or let townsfolk plant little gardens on the ramparts. After the French Revolution, a law in 1791 transferred the ownership of all city fortifications to the state, and they were put under the control of the army. Walks on the ramparts became a thing of the past, the commanders determined when the gates would open and close, and it was they who chose the contractors. To keep the outer sweep of the ramparts free of obstacles, inhabitants were no longer allowed to build within 585 meters of the glacis.[51] From the Napoleonic period onward, the fortifications, which the townsfolk and city councils found to be useful in the eighteenth century, therefore became more and more of a burden to the inhabitants, who no longer had control of them. Instead of a symbol of security and protection, the fortresses and fortifications became a source of unrest and uncertainty, in part because of the unpredictable expropriations and general disruption caused by the work.[52]

---

49 City archives, Oudenaarde (SAO), Nr. 584; 583. Wellington, 'Proclamation', beginning of June 1815.
50 Gils, *De Versterkingen van de Wellingtonbarrière*, pp.20–21.
51 That is: the long, slightly sloping bank around the outer edge of the fortress.
52 Gils, *De Versterkingen van de Wellingtonbarrière*, pp.38–40.

Willem I simply adopted the Napoleonic regulations. In 1816, by royal decree, he declared the plots for the fortresses being built to be state property. Those whose farmland and houses would be appropriated were announced by a special messenger, who 'with the roll of drums', would make the formal announcement somewhere at the village green or marketplace.[53] The residents then had eight days to have their property appraised. A special committee was established to address their grievances and provide compensation, but because the compensation proved to be quite low, the number of citizens appealing the expropriations increased exponentially. One could hardly speak of an 'amicable settlement'. The construction projects did lead to jobs and employment for many, but workers could just as easily be let go when the weather changed for the worse. Their lives were uncertain from one day to the next, and their labour was thankless. Those who had to hoist the Boom clay bricks up the Kezelberg to build the casemates had an especially hard time of it. Some lost fingers to frostbite, others were buried in rubble thanks to sudden landslides. Oudenaarde's chronicler, De Rantere, refers to all kinds of accidents. He also attributes the high number of suicides during these years to the work on the forts; workers would be fired on the spot if the pace slowed or the weather was not cooperating.[54]

It did not take long before collective attempts to subvert these invasions into their daily life were initiated. All kinds of ways were found to engage in more or less open resistance or obstruction. For example, the curate of Oudenaarde's Saint Walburga Church and the city council, citing various prohibitions and regulations, were able to delay the construction of a Protestant church for years. That church was meant to provide the soldiers from the garrison and the officers and engineers from the north the opportunity to hold their own Protestant services. In the meantime, they had made do with a hall in town. Likewise, when a funeral service for a Protestant was held in the hospital chapel, family members were not allowed into the chapel.[55]

The new Dutch subjects also tried to put one over on the 'Hollanders' in other ways. Local contractors were especially skilled in this. The Dutch board had decided to keep the prices low by soliciting bids. Potential contractors could then submit their offers at a price higher or lower than that set by the engineers for moved earth per cubic meter, masonry work, the delivery of stone and brick and the like. By offering a lower price, but in the end also delivering a much lower quality of stone or masonry work than agreed upon, contractors could make a killing.[56] Sometimes the engineering firm also made a deal with the contractor on the spot, bribing them to turn a blind eye and ignore shoddy brickwork and other faults.[57]

---

53 Se, for example, the announced 'roll of the drum', published in the newspaper: 'Vestingwerken der stad Oudenaarde', *Feuille d'Annonces*, Oudenaarde, 9 October 1825.
54 Borremans, *Het Kezelfort*, pp.29–36; De Smet, *Dagboek van Bartolomeus De Rantere*. pp.149–231.
55 City Archive Oudenaerde: SAO HB XVII.A. MOA SN 567.1. Letter Burgerhospital to the mayor of Oudenaarde, 14 June 1826; the governor of East Flanders' response, 3 October 1828; Ordinance Willem I, 7 April 1829. See also Els Vandermeersch-Lantmeters, *Uit de geschiedenis van het Geslacht Vandermeersch te Oudenaarde* (Oudenaarde: Sanderus, 1973), pp.74–75.
56 Various official reports against contractors. SAO HB Nr. 583.
57 See also Uitterhoeve, *Kraijenhoff*, pp.337–339, 345–350.

In addition to the incident in Oudenaarde during Willem I's visit in 1823, mentioned earlier, landslides and collapsing embankments also occurred in Ath and Charleroi. The greatest scandal occurred in Ypres in 1824. After the gunpowder magazine collapsed there, an investigative committee discovered major evidence of widespread corruption. The local supervisor, engineer Lieutenant Colonel Lobry, was sentenced to 20 years in prison; Captain Pasteur, also an engineer, was given one year, and his lieutenant six months. The director of the Third Fortification Directorate, Major General Hennequin, who had been with Willem I in 1823, committed suicide in prison. This affair had another unpleasant consequence, because Baron Krayenhoff, Willem's Inspector General, was also accused. His function was discontinued, and he was put on inactive duty in 1826. Even after being formally rehabilitated and acquitted four years later, his old lustre never returned; he had more or less lost the king's favour.[58] The rumours about malpractice and troubles on the southern frontier were so persistent that even Wellington began to worry. He sent Colonel John T. Jones to inspect matters in the Netherlands, who concluded that indeed, the 'jealous ire of the Belgians' was undermining the construction attempts, further aggravated by the fact that 'by a very impolitic and unnational arrangement, scarcely any but native-born Dutch officers were employed on the new frontier'.[59]

Support for the construction of the fortifications was minimal, and among Belgians, many saw it as a hobbyhorse of Willem I. A cohort close Willem's second son, Frederik, picked up on that dissatisfaction, using it as a means of undermining Krayenhoff's position. Frederik did not see much future in 'all that masonry'. He felt that his father and Krayenhoff had let themselves be pushed too much by Wellington and the British, and in so doing had saddled the kingdom with an overly expensive, unloved and anachronistic military system. With Krayenhoff side-lined, Prince Frederik could bring the somewhat unconventional Fortifications Service and Krayenhoff's corps under his own authority.[60] With that, the future of the allied security project was irrevocably sacrificed on the altar of internal Dutch manoeuvrings.

Whether the fortifications project would have ever become very popular in the Netherlands, aside from Willem and Krayenhoff, is an open question. The variety in the forms and types of obstruction certainly increased their cost. And it was an expense sorely underestimated by Wellington and Willem in the first place. When the fund was dissolved in 1854, the total cost of the barrier proved to have risen to 88.56 million guilders. After deducting the Britain's (one-time) contribution of two million pounds and the French contribution of 60 million francs, the 35.19 million guilders remaining was paid by the Dutch treasury.[61] The Oudenaarde fortress cost 3.4 million guilders, and the expenses for Dendermonde and Ghent were 3.0 and 3.3 million guilders, respectively. Those costs had to be borne by the population.

---

58   Gils, *De Versterkingen van de Wellingtonbarrière*, pp.22–23; Uitterhoeve, *Kraijenhoff*, pp.350–363.
59   See H.D. Jones, *Reports relating to the Re-Establishment of the Fortresses in the Netherlands from 1814 to 1830* (London: Spottiswoode, 1861), pp.ix–xviii, here xv–xvi. See also Jones to Wellington, 28 September 1828, Woolwich, pp.282–283.
60   Uitterhoeve, *Kraijenhoff*, pp.345–357. See also Sterkenburgh, *Van bufferstaat tot neutraliteit*, pp.23–26.
61   Gils, *De Versterkingen van de Wellingtonbarrière*, p.24.

European security had come at a high price; something that Wellington realised as well. After his inspection tour in the summer of 1817, he likewise acknowledged to Castlereagh: 'I am a little afraid of the expense to the King, as the work is going on in really the best style'.[62]

These costs were indeed too high, certainly for the inhabitants of Belgium. The majority were not at all excited about unifying with the northern part of the Netherlands or about the formation of a buffer state. However, the allied security brokers at the time did not pay them much heed. Krayenhoff was the first one to pay the price. But Willem I (and the Allies), too, suffered the consequences when the Belgians revolted in 1830.[63]

What Remained of the Chain of Security

The importance of European cooperation was manifest symbolically in stone and masonry from 1815 on. The new and refurbished fortifications were the most visible result and artefact of that enduring European struggle against terror. They were meant to represent power and protection, explicitly professing the European spirit of order and peace in which they were grounded. Willem I, for example, had the white marble stele removed that Vauban had built into the beautiful façade of the (old) Menin Gate in the city walls around Ypres. The inscription 'Louis XIV' was engraved in gold letters. Willem turned the stone around and had a new text engraved on the backside and put in place in 1822. That inscription reads (in English translation): 'After Europe found peace, and conquered Napoleon, Willem I has reinforced the city of Ypres … Citizens, recently returned under a prosperity rendering government, feel safe and secure, a magnanimous king … devotes himself to your defence. Anno 1820.'[64] In the citadel in Ghent, the role of the Wellington Barrier was emphasised as well. On the monumental entrance gate was chiselled 'Nemo me impune lacesset' – no one attacks me with impunity. And 'Anno xi post proelium ad Waterloo extructa' – erected in the eleventh year after the Battle of Waterloo.[65] Whether the people of Belgium actually 'felt safe' with those fortresses, as the inscription on the Menin Gate proclaimed, was the question.

By the end of the 1820s, most of the fortifications had been rebuilt, but less than two years later, in 1830, the Belgians revolted against the – in their eyes – harsh and unjust rule of Willem I, which systematically favoured Protestants and the north. After a few military battles and years of negotiations, the Belgians, with the support

---

62 Wellington to Castlereagh, 7 August 1817, WSD, vol.12, p.23.
63 Nelson, 'The Duke of Wellington and the Barrier Fortresses after Waterloo', p.42.
64 Commemorative stone from the old Menin Gate, which was torn down after Belgium became independent; only the stele remains, and can still be seen from the Rijsselpoort (or Lille Gate). Latin text: 'Pacata Europa subverso Napoleonte Gulielmus I Urbem Iprensem olim vale munieam a Ludovico XIV. Validioribus propugnculis cirxlmdatam novis denvo suppressis allies niumittonibus restituit civs felici imperio neiper restitu sicupupi estote. Rex magnanimous consilio sagm animo fortis labore indefessus. Incolumitati vestræ toto iectore incumbit. Anno MDCCCXX'.
65 Gils, *De Versterkingen van de Wellingtonbarrière*, p.65.

of the Allies, managed to force the separation of the Southern Netherlands and to establish their own kingdom. With that, the Dutch buffer state, the Boulevard of Europe, was split in two. The Wellington Barrier came to lie in Belgium, but the citizens there were more concerned with industrialization and modernisation than with defence measures against France. In 1839, the first thing the new king Leopold decided was to dismantle the fortifications along the barrier. The old Menin Gate in Ypres, for example, gave way to city expansion; at three meters, the passageway was too narrow and the recurrent cause of traffic jams.[66]

From a military perspective, the fortresses were also a cause for concern. Maintaining them was too expensive, and the Belgian army was too small to occupy and defend them on its own. The Treaty of London of 1839 had also stipulated that Belgium was to remain 'perpetually neutral'. Preventive occupational forces coming from Prussia and Great Britain – an agreement that Willem I had painstakingly negotiated in 1818 – were, therefore, no longer an option. The foremost line of the defensive barrier –Ypres, Menin, Ath, Philippeville and Marienbourge – was the first to go. The expensive, skilfully planned fortifications with their complex ravelins and strongholds were, for the most part, demolished. The first railway lines finished the job. Only the National Redoubt around Antwerp remained.[67] It functioned as a defensive post to which the army could retreat and regroup, awaiting support from the great powers that stood behind Belgium's independence. After that, the second ring of fortifications was also demolished: Ostend, Nieuwpoort, Oudenaarde, Mons, Charleroi, Dinant and Huy. After 1870, the citadels of Ghent and Dendermonde had to make way for progress: for a city park, and later in 1913 for the *Feestpaleis* that was built on the grounds of the Ghent citadel on the occasion of the World Expo.[68]

Only a few fortresses along the Meuse remained of the Wellington Barrier, and a commemorative plaque or two. Images of the Dutch coats of arms were replaced by Belgium's.[69] In 1914, the ramparts of a number of fortresses, including Dendermonde, were actually used as a defence. Thereafter, the Germans demolished them. After the First World War, what remained of the fortifications was demolished – there was not a bastion or bulwark still intact.[70] What did remain were the underground networks of passageways, about which all kinds of rumours and legends continued to circulate. For example, some underground tunnels would extend far beyond the fortress and end up in a nunnery. Most of what remains of the fortifications is no longer visible to the naked eye, nor do they enjoy legal protection. From all the large fortresses, bastions and bulwarks, hardly one remained intact. The Kezelfort in Oudenaarde was the only stronghold that survived, partly because it was and is

---

66 The Ypres fortresses were demolished around 1850; the gate was also dismantled, and passageway 13 meters wide was put in its place. The commemorative stone was placed on the Lille Gate.
67 See A. Picon, 'Die Ingenieure des Corps des Ponts et Chaussées.Von der Eroberung des nationalen Raumes zur Raumordnung', in H. Stück and A. Grelon (eds.), *Ingenieure in Frankreich, 1747-1990* (Francfort/New-York: Campus, 1994), pp.77-99.
68 Gils, *De Versterkingen van de Wellingtonbarrière*, pp.43, 59, 65. The Redoubt was completed in 1859.
69 Gils, *De Versterkingen van de Wellingtonbarrière*, pp.24-26, 65.
70 Gils, *De Versterkingen van de Wellingtonbarrière*, p.51.

owned by a private family, and partly because of its location as a resting place for bats and therefore not open to the public.[71]

Reading the Landscape of Security

At the Congress of Aix-la-Chapelle in September 1818, two years earlier than planned, Britain, Austria, Prussia, and Russia decided to withdraw the occupying forces from France, and by 30 November, the evacuation was complete. With that, the allied security order was declared 'finished' for the time being. The debts were paid, the provisions of the second Treaty of Paris had been met and the great powers could turn their attention to new projects of trans-imperialist cooperation. The significance of the Congress actually lay in establishing the translation of the elusive and ambiguous imagery of 'the balance of power' into the practice of constructing forts, which were to represent the power of the Quadruple Alliance in Europe. They were also the only concrete point of the allied dialogue that was agreed upon as a multinational security project in 1818 and recorded (albeit in a secret protocol). The forts were thus also a visible starting point for criticising the Allies' dominance.

Bricks (forts), earth (defensive ramparts) and paper (treaties and interest-bearing securities) had to hold the allied peace and security in concert. The episode in Oudenaarde, however, has demonstrated that not everything constructed was equally solid. Moreover, the artefacts of collective security measures did not always live up to the repute of allied power. How life was perceived along the border with France was different in 1830 than it was in 1815–1818. France was no longer considered an enemy or aggressor. The bulwark of Europe, as a network of fortifications, had lost its function. As those visible manifestations of collective security deteriorated, the reputation of allied solidarity also crumbled. At the same time, the instruments and artefacts that were supposed to curb or restrain the terror had reawakened or spawned new protests to the plan. In the eyes of the courts of Spain and Portugal, of the Belgian population, of German students, of liberals and radicals spread across the continent and in South America, the four great powers had erected a far too exclusive and imperialist security policy. At best, these 'objects' of the European security system shrugged their shoulders and, when it came right down to it, took little notice of the Allied artefacts of strength. In the worst cases, these demonstrations of allied power, however, inspired new waves of resistance and rebellion.

---

71 Gils, *De Versterkingen van de Wellingtonbarrière*, pp.76–78.

4

The Logistics of a Successful Siege: The Planning for the Allied Siege of Ciudad Rodrigo, January 1812

Mark S. Thompson

Background

In March 1811, Wellington was waiting for the French forces that had been in front of Lisbon since October 1810 to face the inevitable and retire. Having spent a few weeks in front of the Lines of Torres Vedras, Masséna's French army had retired to around Santarém to wait for reinforcements or new orders from Napoleon. By the last days of February 1811, it was clear that there was no option but to retreat. Whilst the allied forces still held the southern route into Portugal through Badajoz and Elvas, Wellington's priority would be to recapture the northern route through the fortresses of Ciudad Rodrigo and Almeida, both of which had been captured by the French in summer 1810. Control of the two routes into Portugal was essential before any thought could be given to advancing into Spain.

Unfortunately, Wellington's plans were upset by the French capture of Badajoz in March 1811, just as the pursuit of Masséna into northern Portugal had commenced. A scramble now ensued to put together troops and guns to retake Badajoz. The task was given to *Marechal* William Carr Beresford, who moved his troops south and worked to collect enough guns by taking them from the Portuguese fortress of Elvas. The siege train that was collected was inadequate in terms of the number of guns and their quality, and the two attempts at besieging Badajoz made no progress before they were both interrupted by advancing French forces.

Many commentators have criticised Wellington for not despatching a suitable siege train from Lisbon for this purpose. As this paper will show, moving a siege train and its equipment was a huge task. There was no way a siege train could be collected and moved across Portugal from Lisbon in the available timescale.

Even while Wellington attempted to retake Badajoz, detailed planning for the more important task of recapturing Ciudad Rodrigo was underway. This paper will describe the enormous and complex preparations that were needed to move a large siege train, its ammunition and stores across 400km of difficult terrain.

The main tasks can be summarised as follows:

1. Sourcing the guns. Ideally, these needed to be new iron guns. Older guns would be worn, and this significantly affected both power and accuracy. Brass guns overheated with constant use.
2. Sourcing powder and shot for the guns. A single siege gun could fire 100 shots a day, requiring 1,500kg of materials to be moved.
3. Sourcing the road or river transport to move the guns. Carts were in short supply throughout the Peninsular War, and getting together the large number needed to move a siege train was immensely difficult. The animals and civilians employed to move the materials needed to be fed and paid.
4. Planning the route. You could not rely on maps for planning a route. Many roads in the Peninsula were of poor quality or had blockages that restricted the movement of artillery. Often, repairs had to be made to roads to make them passable. Bad weather would quickly make roads unusable. Similarly, rivers needed to be surveyed and blockages removed. The condition of a river in different seasons could affect the ability to use it.
5. Sourcing the specialists to operate the guns. Whilst artillerymen from the field units were often used, they sometimes needed to be supported by additional artillerymen (or sailors) brought in specifically for the siege.
6. Sourcing troops trained in siege work. Few were available until 1813, so troops from the Line regiments needed to be trained.

The key learning point is that the bulk of the planning and effort for any siege occurred in the period prior to the first gun being fired. The actual battering of the walls was the easy bit.

The Basics of a Siege: Equipment, Defence and Attack.

Equipment Needed for a Siege

There were three basic types of gun, each of which had a different role: the cannon, the howitzer and the mortar. The cannon was the most powerful and fired directly at a target. It could use solid iron shot or explosive ammunition. To breach a wall required round shot. A howitzer, with a short barrel, was designed to fire at higher trajectories over obstacles. Typically, they used some form of explosive ammunition. Their role was primarily to shoot at people. In some instances, they could fire directly at targets, but their range and accuracy were less than a cannon. A mortar was a weapon that was also designed to fire over objects. Typically, they fired heavy explosive shells to injure people and buildings. Howitzers and mortars had little ability to breach the walls of defensive places.

Cannon and howitzers usually travelled on wheeled carriages, whilst mortars did not and required separate wheeled transport. The larger siege guns could weigh up to 4,000kg. On rough roads, they would break their carriages. On wet roads, they were difficult to move through the mud. Where possible, the heavier cannon would travel long distances on separate carriages to avoid damage to the carriage that would be used to fire the weapon.

Moving a Siege Train

Before the actual movement of equipment for the siege of Ciudad Rodrigo is described, it will be useful to give an outline of the scale of the task to move a battering train and its ammunition.

Major Alexander Dickson RA estimated that to move a 24-pounder cannon, up to 12 oxen were required in good weather and up to 16 oxen in bad weather. He specifically reported that eight pairs of oxen were used for the siege of Ciudad Rodrigo.[1] To transport two day's worth of shot and powder would require 12 carts with 24 oxen and a minimum of 15 civilians.[2] To move a substantial siege train of 40 large guns with two days' supply would therefore require 1,600 oxen, 500 carts and 600 civilians. The total weight being moved was nearly 300 metric tonnes. The convoy would be 10 kilometres long and would move 10–15 kilometres daily in good weather. At the end of each day, the tail of the convoy would get to where the head of the convoy started. These figures do not include the transportation of food for the animals, soldiers, and civilians involved in moving the siege train.

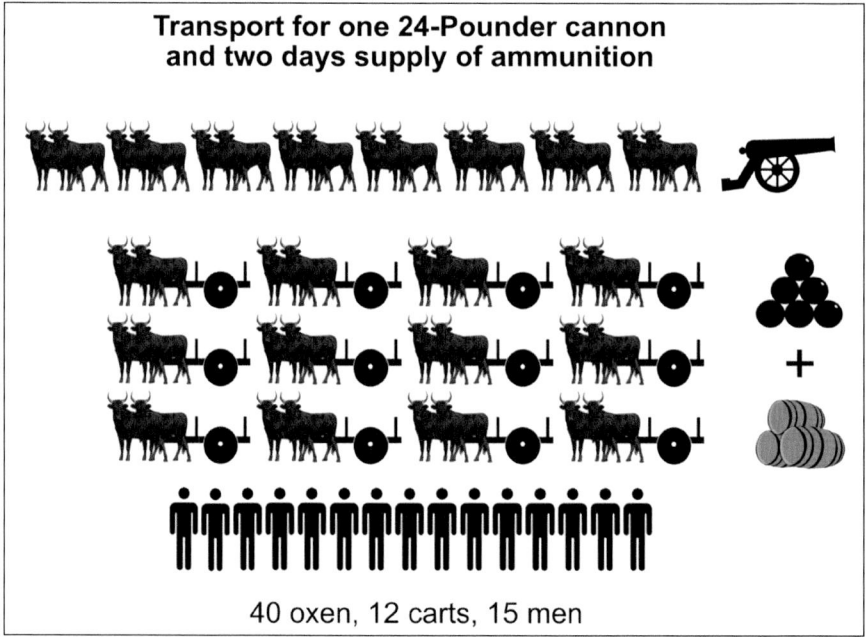

40 oxen, 12 carts, 15 men

---

1   John H. Leslie (ed.), *The Dickson Manuscripts, Being the Diaries, Letters, Maps, Account Books, with Various Other Papers of the Late Major General Sir Alexander Dickson* (Cambridge: Ken Trotman, 1991), vol.3, p.477.
2   Estimated at 100 round shot per day and eight pounds (3.6kg) of gunpowder per shot for a 24-pounder cannon.

Apart from the pure numbers, there were other significant issues with transport. The animals and civilians mentioned above were all hired locally for the task. Generally, they were not supplied with food for themselves or their animals, and their pay was intermittent and always significantly in arrears. The civilians became more desperate as they moved further from their homes, and desertion was a significant issue. On occasion, guards were placed on the civilians to stop them from absconding, and this made their situation even worse as they were unable to forage for food even if any was available in the wake of a travelling army.

The other key issue was availability. Collecting hundreds of carts and oxen in areas which had already been swept of available transport was incredibly difficult. This would require early planning to avoid taking the transport from the army and commissary, which would almost certainly affect the movement of food.

The Planning for the Siege

Roads and Rivers in the Peninsula for Transport and as Barriers
The Iberian Peninsula is an area dissected by large rivers and mountains. These features dictate the direction and quality of roads. Much of the land had few occupants, and what roads there were connected the major conurbations. There were good 'postal' roads in Portugal between Lisbon, Coimbra, Porto and Elvas. In Spain, similar quality roads existed between Madrid and Cadiz, Cartagena, Barcelona (to France), Santander (via Valladolid and Burgos), San Sebastián (to France) and Badajoz (and on to Lisbon). Lesser quality roads would connect other towns and cities, and their quality would deteriorate quickly with heavy use as few were paved. The routes of the roads were all dictated by the mountains and the availability of crossing points for the major rivers. Many rivers were impassable for some or all of the year without bridges. Moving heavy artillery required access to good roads and bridges, and therefore limited available routes. From Lisbon to Ciudad Rodrigo, there were very few possible routes, and this determined how and how quickly Wellington could move his siege train. Often, bridges were near major towns and would not be accessible if the enemy held the place.

The early years of the war saw Wellington despatching officers to survey routes across the Peninsula, building an understanding of the routes available to the different parts of the army. This was usually supplemented by specific surveys prior to the movement of siege trains. For example, Dickson rode across the likely routes for the advance of the siege train for Ciudad Rodrigo in the summer of 1810, and adjustments were made to the route based on his information.

The Collection and Delivery of the Siege Train and Siege Materials
The first indication in Wellington's correspondence of his intention to besiege Ciudad Rodrigo came on 14 May 1811 when he ordered the new siege train recently arrived at Lisbon to be transported to Porto.[3] This was a full eight months before Ciudad Rodrigo

---

3   J.G. Gurwood (ed.), *The Dispatches of Field Marshal the Duke of Wellington, KG, during His Various Campaigns ... from 1799 to 1818.* New Edition (London: John Murray, 1837), vol.7,

was taken. On this date, Beresford was undertaking the first siege of Badajoz and was just suspending it to fight the Battle of Albuera to stop the French under *Maréchal* Jean Soult relieving the fortress. Wellington was still in northern Portugal, having recently fought the Battle of Fuentes d'Onoro to stop the French relieving Almeida.

This siege train was made up of thirty-four 24-pounder and four 18-pounder cannon, eight 10-inch iron mortars, two 8-inch brass howitzers, twenty 24-pounder iron howitzers and ten 5½ inch mortars, totalling 78 pieces of ordnance.

On 19 July 1811, Wellington, who was still near Elvas, called together his commissary, artillery and engineering commanders and issued instructions to move the siege train from Porto towards Ciudad Rodrigo.[4] The plan was in three stages. First, to move it by boat from Porto to Lamego up the Douro river. Second, to move the train by land from Lamego to Trancoso and third, to move it forward to the vicinity of Ciudad Rodrigo. These original instructions were altered later, but they will be described as originally written.

In the memorandum, Dickson was ordered to Porto to superintend the move of the artillery and ammunition and an engineer officer, Lieutenant Anthony Marshall, was appointed to manage the engineering stores.[5] On the same day, Wellington ordered two companies of Royal Artillery (Bredin's and Glubb's) to be moved from Lisbon to Porto to assist Dickson's with the transfer of the equipment.[6]

Although Wellington's memoranda were typically precise in what was required, there were clearly still a number of uncertainties. Dickson, although ordered to Porto, was asked to go via Almeida to see what ammunition could be provided locally. Transporting round shot was more resource-intensive than transporting the cannon. Similarly, powder was also an issue. Wellington, writing to Beresford, noted that they had half the required amount at Porto with the siege train but complained that he could not get reliable information on how much was available at Lisbon. He emphasised that 1,600 barrels needed to be sent to Porto as soon as possible. He clearly saw this is a priority. Options he gave Beresford included taking powder from the Lines of Torres Vedras, borrowing it from Cadiz (for which he had written to his brother to arrange transport in case that option was required), or stationing a ship off Porto to divert powder ordered from England.[7]

There appears to be an urgency about Wellington's actions at this time. Dickson was tasked to have the siege train at Ciudad Rodrigo before the end of September. Wellington wanted substantial powder reserves available at the same time. Training of soldiers from the Line regiments for siege work was also commenced in July. It looks like Wellington was intending to make a dash at Ciudad Rodrigo in the autumn, and that's why he was pushing for the material to be ready. Wellington wrote: 'In consequence of reports which I received, that the garrison of Ciudad

---

    p.552, Wellington to Howarth, Villar Formoso, 14 May 1811.
4   Gurwood, *Dispatches*, vol.8, pp.121–122, Memorandum for Colonels Framingham and Fletcher and Mr Kennedy, 19 July 1811.
5   Royal Engineers Museum (REM): 5501-59-1, Jones MS Journal, vol.2.
6   Gurwood, *Dispatches*, vol.8, p.122, Wellington to Peacocke, 19 July 1811.
7   Gurwood, *Dispatches*, vol.8, pp.125–129, Wellington to Beresford; Wellington to H. Wellesley; Wellington to Liverpool, all 20 July 1811.

Rodrigo were in want of provisions, which reports were not contradicted in time, I have brought the army to this quarter sooner than I intended.'[8]

Any thought of an early attack was stopped by the French successfully resupplying Ciudad Rodrigo in late September. *Maréchal* Auguste de Marmont had proved he could bring forward substantial forces to support the town. Attempting a siege under these circumstances would be very risky.

A Breakdown of Wellington's Memorandum

Let us take a look at Wellington's memorandum of 19 July 1811. This was still six months before the actual siege started. The initial proposal was to move only 48 guns to Ciudad Rodrigo. These are shown in Table 2.

Table 2: Planned siege train for Ciudad Rodrigo

Gun	Number	Ammunition (each gun)	Ammunition (total)
24-pounder cannon	34	350	11,900
18-pounder cannon	4	350	1,400
10-inch mortar	8	160	1,280
8-inch howitzer	2	160	320
Total	48		14,900

With the guns were 350 rounds for each cannon and 160 rounds for each of the mortars and howitzers. The total weight to be moved was about 425 metric tonnes, the cannon being about 160 tonnes. To move the remaining 265 tonnes, the largest land vehicle available could carry about 300kg.

Wellington did not detail the shipping requirements from Porto to Peso da Régua, the nearest dock to Lamego. Dickson and Richard Kennedy, the Commissary General, were to arrange them.

Once the equipment was at Peso da Régua, 384 pairs of oxen would move the guns. Royal Artillery officer Captain John May wrote that eight pairs would be used for the 24-pounder cannon.[9] Less would have been needed for the mortars and howitzers, perhaps three or four pairs. The remainder would be a reserve of animals.

To move the stores, Wellington ordered 892 country carts which were distributed as shown in Table 3.

Table 3: Loads on carts

Equipment	Carts
1,200 barrels of powder (7 per cart)	171
24-pounder shot (350 per gun)	396
18-pounder shot (350 per gun)	35
8-inch & 10-inch shells	200
Laboratory and general stores	90
Total	892

---

8   Gurwood, *Dispatches*, vol.8, p.188, Wellington to Dickson, 13 August 1811.
9   John Jones, *Journal of the Sieges Carried on by the Army under the Duke of Wellington in Spain During the Years 1811 to 1814.* 3rd ed. (London: John Weale, 1846), vol.1, p.358.

These 892 carts would be required to make multiple trips to move all the equipment. An initial two trips would move all the ammunition to Lamego. A further two trips would then be required to move the ammunition from Lamego to Ciudad Rodrigo. The various descriptions are quite confusing in describing the movements of the carts, and it is easier to think in terms of cart journeys rather than physical carts. So, 1,784 cart journeys would be needed to move the ammunition to Lamego and then a further 1,784 cart journeys to move the ammunition to Ciudad Rodrigo. An additional 200 carts were requested to carry the engineer stores, making a total of 3,768 cart journeys.

In summary, Wellington's memorandum of 19 July detailed 384 pairs of oxen and 1,092 carts to make the initial move of equipment. This was 2,952 oxen and 1,092 carts. Accompanying them would be 1,200–1,500 civilians, 200–300 artillerymen, plus whatever guard was deemed necessary.

What this huge endeavour would deliver was the guns and two days' supply of ammunition. To complete the siege, a great deal more powder, shot and shell would need to be delivered. The Commanding Royal Engineer, Lieutenant Colonel Richard Fletcher, estimated usage for the siege would be 25,000 round shot, which was about double what was moving with the guns.[10]

The Movement of the Train for the Siege of Ciudad Rodrigo

It was subsequently decided to take the 24-pounder iron howitzers. These additional guns would have increased the demand by another 80 oxen pairs for the guns and 125 cart journeys for ammunition.

Dickson was ordered to Porto via Almeida, arriving on 28 July, and his initial impression was that it presented a 'woeful aspect'. The fortifications were in ruins after the explosion at the end of the French siege in 1810 and further damage when the French abandoned it in May 1811. Dickson also noted meeting Captain George Macleod RE, who had been sent there to report on the state of the defences.[11] MacLeod's report was not encouraging: 'Of the six fronts, four are almost rendered defenceless … the bridge and gate leading from the ravelin to the body of the place totally destroyed'. He concluded that basic repairs could provide limited defence and would require a working party of 500 men for a month.[12]

Dickson reviewed the available round shot and concluded there were about 4,000 Spanish 16-pounder shot that would be suitable for the British 18-pounder cannon. The 24-pounder shot were a mix of different calibres, and Dickson asked the governor to arrange for them to be 'gauged and to select those of the proper calibre'. Dickson, writing to Wellington a few days later, was pessimistic about the number that might be found.[13]

---

10   Jones, *Sieges*, vol.1, p.358. Jones mentions the daily fire rate and estimate for the whole siege on this page. Fletcher's estimate for the siege used a lower daily fire rate.
11   Leslie, *Manuscripts*, vol.3, p.426.
12   REM: 2001-149-11, Report on Almeida, 20 July 1811.
13   Leslie, *Dickson Manuscripts*, vol.3, pp.429, 434.

Dickson then travelled on to Trancoso and was concerned about the quality of the roads and the crossing points over the river Côa. The main bridge over the Côa at Almeida was broken, and he recommended that the ford just north of the bridge was suitable if the roads on either side were repaired. He noted the road from Trancoso to Moimenta da Beira was 'execrable, and not fit for the movement of heavy artillery'. Dickson wrote to Wellington on 1 August and recommended 'the train march directly to Pinhel'.[14] On his first day in Lamego he met with *Tenente-general* Bacellar, the Governor of Beira province, and arranged for the roads to Pinhel to be repaired and for an escort of 1,000 Portuguese militia to be provided for the siege train. Dickson now travelled down the river Douro to Porto, arriving on 3 August. The two Royal Artillery companies of Bredin and Glubb had arrived the day before. Unloading of the transports started the following morning.

The movement of the siege train up the river now faced another challenge, which the port wine companies had faced for many years. The river was not smooth; it contained obstructions, rapids and shallows. Large boats could not get through the shallow areas, and small boats could not get up the rapids. It was not unusual to have to unload and reload cargo when traversing difficult areas. John Jones left a description of the method of a boat travelling upriver:

> We passed many boats of 36 pipes [medium sized] towing against the stream by 16 or 18 men and one boat had the addition of a pair of oxen. Although it was blowing a strong gale directly aft and the boat had numerous large square sails set they did not appear to advance two miles an hour. Particularly we observed an excellent towing path on the right bank of the river, but generally the men employed to tow were obliged to pick their way over rugged and sharp rocks and in various few parts it seemed utterly impossible for them to walk in any manner from the perpendicularity of the rocky bank to the edge of the water.[15]

Dickson reported that, it being the height of summer, the largest boats could not be used at all due to the water level. The main problem area was at the rapids above Vimieiro. Here, the small boats, the *Trafegueira* could not safely pass. The medium sized boats, the *Matriz*, could pass but only with partial loads. The solution was for a mix of both sized boats to move upriver to Vimieiro, and once the partly loaded *Matriz* had climbed the rapids, the cargo from the *Trafegueira* would be transferred onto the larger boats. Under these arrangements, boats were despatched as described in Table 4.[16]

---

14   Leslie, *Dickson Manuscripts*, vol.3, pp.432–4.
15   REM: 5501–59–2, entry for 5 December 1811.
16   Jones, *Sieges*, vol.1, p.359.

Table 4: Boats used to move siege train up Douro

Date	*Matriz* (larger)	*Trafegueira* (smaller)
7 August 1811	21	19
8 August	19	18
9 August	26	13
10 August	12	2
11 August	4	3
12 August	9	15
13 August	7	10
14 August	7	9
Total	105	89

With the upriver journey expected to take 10 days, the larger boats would not return for about two weeks. The small boats, only going part-way, would return quicker.

Dickson wrote to Wellington on 9 August, stating that all the guns and much of the ammunition had been despatched, but there was still more to transport. The availability of river boats had reduced, but the first of the smaller boats had returned. He also reported that the ship from Lisbon carrying 1,600 barrels of powder had still not arrived.[17] A second letter, four days later, reported that there was still no powder ship. Of more immediate concern to him was that he had not had a reply from Wellington about changing the route of the siege train. He really needed to know this before he started planning the route from Lamego. Dickson also reported that he had received an answer from Almeida, and there were only 2,100 24-pounder shot available, far fewer than had been hoped.[18]

Whilst the first elements of the siege train were being moved up the river, Dickson was arranging for alterations to be made. The siege train was constructed on the assumption that it would be pulled by horses. There were few horses available, so there was no alternative to using oxen. Unfortunately, oxen required different fittings to pull the guns. A horse was fitted with a harness that connected to a pole on the limber. Oxen had a yoke which fitted over the beasts' shoulders. He reported on 5 August that he was having 228 *solinhos* (yokes) made to fit to the gun limbers. The limbers would be the last of the equipment to be moved up the river, when the changes had been made. Also, to transport the round shot, he arranged for 631 boxes to be made to fit on the open-sided country carts. Writing to Wellington on 6 August, he estimated that they would take eight or nine days to complete, although another letter three days later reported that the work on the yokes was going very slowly and he had decided not to make them for the 24-pounder howitzers, significantly reducing the number required.[19] In the same letter, Dickson expressed concern that there was a shortage of travelling carriages for the guns and mortars. He noted: 'There is no remedy but for the guns to go on their own carriages, and I

---
17 Leslie, *Dickson Manuscripts*, vol.3, p.440.
18 Leslie, *Dickson Manuscripts*, vol.3, pp.441–442. Eventually, nearly 8,000 were found.
19 Leslie, *Dickson Manuscripts*, vol.3, pp.437–438.

dread much the consequences of this, that the carriages with the guns mounted on them going over such [bad] roads as they must pass, will be so much shook as to be rendered unfit for the service required of them.'[20]

There was a lot of work needed before the new siege train could begin its journey.

The first artillery officer set off for Lamego on 11 August to start receiving the equipment when it arrived. By 13 August, nearly 300 boxes and 80 *solinhos* had been made, with Dickson estimating that they should all be complete by 17 August. The non-arrival of the powder ship from Lisbon was now causing serious concern as the transport upriver would be complete before it arrived, requiring boats and people to be held back until it arrived.[21] By 14 August, all the other goods had been shipped and attention moved to Lamego and the onward movement of the siege train. Unfortunately, no decision had been made on the route to be taken.

On 26 July, Wellington left Elvas to travel north, not arriving until about 12 August. It would appear that some of Dickson's correspondence was delayed over this period as it tried to follow Wellington, who was moving 300km north. Wellington, writing to Dickson from Fuenteguinaldo, 25km south-west of Ciudad Rodrigo on 13 August, reported that he had not received Dickson's letter of 1 August, although he had received later ones. The missing letter was the one in which Dickson asked Wellington to approve the change of route to avoid the bad road to Trancoso. It was not until 17 August that Wellington received this letter and replied; it would have taken a few more days for Dickson to receive the reply. He approved Dickson's recommendation to avoid Trancoso and asked him to nominate another concentration point for the siege train. Wellington also confirmed that he had ordered repairs to be made to the roads at the ford over the Côa near Almeida.[22]

Just to further complicate matters at this time, Dickson had fallen ill on 14 August, and it was not until 25 August that responsibility was officially handed over to Major John May RA. This probably had only a limited short-term impact as May was already involved in the process, and the other officers knew what had to be done. Assisting May were the two companies of Royal Artillery and 100 Portuguese artillerymen. May also made an urgent request for wheelers and blacksmiths to maintain and repair the gun carriages as the 'wheels of the 24 prs are [already] much shook'. Dickson added that the guns had been on the ship for three years, and the wheels would require 'drawing together'. Having done that, he expected no further issues with the wheels.[23]

The troops to support the siege train were increased. Bredin's company was replaced by Holcombe's company, and the number of Portuguese artillerymen was increased to over 300. In addition, nearly 1,500 Portuguese militia were allocated to guard and to help move the guns; most of these were later replaced with Ordenança.[24]

---

20   The terminology used is not consistent. I suggest that Dickson is complaining about the lack of sling type carriages for transporting the guns. Using different carriages for transporting the guns over long distances on poor roads would reduce the damage to the primary gun carriages.
21   Leslie, *Dickson Manuscripts*, vol.3, p.441.
22   Leslie, *Dickson Manuscripts*, vol.3, p.445.
23   Leslie, *Dickson Manuscripts*, vol.3, pp.452, 469.
24   Leslie, *Dickson Manuscripts*, vol.3, pp.453, 459, 460.

Whilst moving the siege train 70km in five weeks does not seem that impressive, other one-off factors need to be taken into account. These included altering the limbers to be pulled by oxen, preparing boxes to carry the round shot, moving artillery companies to manage the guns, and collecting nearly 200 boats, 1,100 country carts and an additional 768 oxen. The preparations were slow, but the siege train was now ready to move, except most of the powder for the guns had not arrived!

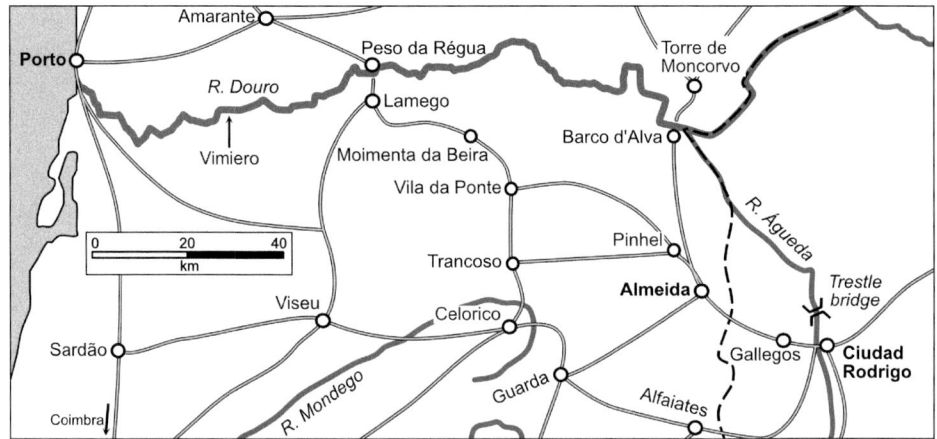

The route of the siege train to Almeida.

As the guns arrived at Peso da Régua they were quickly forwarded to Vila da Ponte. Dickson noted that fourteen 24-pounders and some stores were at Vila da Ponte by 25 August. Somewhat confusingly, there is an 'Instruction' dated 26 August, splitting the siege train 'moving from Lamego' into divisions. This was likely to be an organisational decision to split responsibility for the guns into manageable groups rather than the order to start the move. The instructions also required 350 rounds for each cannon and 160 rounds for each howitzer and mortar to travel with the guns. The plan was to increase the ammunition to 800 rounds for each cannon and 400 rounds for the other ordnance, and Dickson reported on 27 September that this had been achieved. The weight of the ammunition alone was over 500 metric tonnes and would require about 2,000 carts.[25]

On 3 September, Dickson was still complaining that the powder had not arrived. He noted that the ship bringing 1,600 barrels of powder had sprung a leak and returned to Lisbon.[26] When the powder finally arrived at Peso da Régua, as they had no tarpaulins to cover it, Dickson left it there until covers could be made. The weather, particularly through October, was very wet.[27] A further mistake in ordering the tarpaulins meant that the powder was still not moved forward in December 1811 when Wellington's order to move the siege train to Almeida was received.

---

25   Leslie, *Dickson Manuscripts*, vol.3, pp.446, 448, 450, 468.
26   Leslie, *Dickson Manuscripts*, vol.3, p.453.
27   Leslie, *Dickson Manuscripts*, vol.3, pp.483,488.

Collecting this amount of transport together was a major task, and keeping the carts and bullocks together for an extended period leading up to the siege was impossible. The scale of this operation was even larger than the numbers above suggest. John May's report of 28 August 1811 identified 4,170 journeys by oxen pairs to move the siege and engineering material forward to Vila de Ponte.[28] There were 1,910 ox carts and 4,920 oxen physically present with the siege train at Vila de Ponte at the end of September 1811.[29] This huge number could not be sustained as the commissariat had been stripped of transport to achieve the move. Many of the carts and animals were returned to allow routine supplies to the army to continue. This would create difficulties for the move to Almeida and major issues when the siege was eventually ordered.

Transport for the mortars remained a problem, and they were still at Lamego on 6 September, all the other guns having already moved forward. Dickson commented that some were sent on overloaded 'common bullock cars, which I fear will not stand the march'. They did eventually make it to Vila da Ponte, where new 'strong' carriages were made for them.[30]

The siege train that collected at Vila da Ponte is detailed in Table 5.

Table 5: Siege train at Vila de Ponte

Gun	Number	Rounds per gun
24-pounder cannon	34	800
18-pounder cannon	4	800
10-inch iron mortar	8	400
8-inch brass howitzer	2	400
5.5-inch (24-pdr) iron howitzer	16	400
Total	64	

Twenty 24-pounder howitzers had been shipped, but Wellington had diverted four of them to replace the lighter howitzers in the 9-pounder foot artillery brigades.

Although the siege train was at Vila de Ponte by mid-September, it did not move forward for two months. Dickson originally expected that the train would move on quickly, but when Wellington ordered much of the transport back to Lamego to move stores for the army, he realised that the initial urgency had dissipated. Wellington had realised that the information he had been given that Ciudad Rodrigo was low on stores was wrong. There was also concern towards the end of September as it became clear that the French were planning to advance on Ciudad Rodrigo. Whilst the expectation was that the French were just planning to resupply

---

28 Jones, *Sieges*, vol.1, table facing p.363. The report is mentioned in Leslie, *Dickson Manuscripts*, vol.3, p.456 but not included. May reported on the number of oxen journeys required to move the siege train forward, not the number of oxen present. In summary, 2,071 oxen pairs for the first trip; 1,367 for the second trip and 732 to increase the ammunition up to 800 rounds for cannon and 400 for howitzer; making 4,170 oxen pair journeys in total.
29 Jones, *Sieges*, vol.1, p.364.
30 Leslie, *Dickson Manuscripts*, vol.3, pp.455, 458, 469.

the fortress, there was a chance that a more general advance was being planned, and Wellington did consider moving the siege train back to Peso da Régua.[31] There were a few anxious days before it became clear that having resupplied Ciudad Rodrigo in late September, the French were retiring.

In early October, Dickson commented on concerns about forage for the many animals. He noted that the engineer animals were suffering as they were all being kept close to the stores, but he was foraging animals up to 15 kilometres away and was having no problem feeding them.[32]

During this period, work was done to repair the gun carriages and improve the roads in the area. As late as the end of October, Dickson had not received any information from Wellington that a move to Almeida was even planned. He commented on 25 October that he had 'received no intimation of that kind, or even hints from Head Quarters'.[33] Much of the defences of the town had been destroyed by the French when they abandoned it in May 1811. Wellington was not willing to risk the siege train until the town was secure. The successful French resupply of Ciudad Rodrigo had demonstrated that Almeida could be at risk from another strong French advance. To speed up the repairs, Wellington ordered a substantial working party of Portuguese militia and allied soldiers to assist. Dickson was also ordered to repair over 90 guns that the French had spiked and left at the fortress, but this work was not started until after the siege guns arrived and took many weeks to complete.[34]

Finally, on 14 November 1811, Wellington was confident that Almeida was secure enough and ordered the siege train forward to Almeida. Wellington thought that many of the oxen to move the guns were around Lamego and would need to be collected.[35] When Wellington's order arrived on 16 November, Dickson immediately replied, informing Wellington that the oxen for the guns had been kept close to Vila da Ponte. He noted that 200 carts with the engineer stores would depart on 18 November and the guns would march 'successively as they [the oxen] arrive'. Dickson then reported that there were no carts to move the stores (ammunition, powder, etc) other than the 200 engineer carts and moving the stores with only these would take a very long time. Dickson estimated there were 3,000 cartloads, excluding the 1,600 barrels of powder 140km away at Peso da Régua, which would require another 250 cartloads. Dickson also expressed concern that any forage for the animals on the route to Almeida would be quickly stripped, and they could starve without additional help. Of more concern, after the first two round trips, the 200 carts had been reduced to 129 through desertion and breakages. Nearly a third of the carts had been lost in two weeks. The wastage would only get worse. The flaws in the system of using civilian contractors for transport without proper processes for pay and feeding became even more apparent. Dickson pleaded that the commissary should make arrangements to send money from Lamego, as he could not wait for the civilians to travel from Vila Da Ponte to Lamego to get the pay.[36]

---

31 Leslie, *Dickson Manuscripts*, vol.3, p.468.
32 Leslie, *Dickson Manuscripts*, vol.3, p.479.
33 Leslie, *Dickson Manuscripts*, vol.3, p.490.
34 Leslie, *Dickson Manuscripts*, vol.3, p.510.
35 Leslie, *Dickson Manuscripts*, vol.3, p.501.
36 Leslie, *Dickson Manuscripts*, vol.3, pp.505, 511, 514.

Another letter from Dickson on 18 November confirmed that with help from the commissary, he had been able to provide about 1,100 oxen for the guns, and the first two divisions of artillery would move that day with the following four divisions at daily intervals. Table 6 summarises the divisions of the siege train.[37]

Table 6: Division of guns for move to Almeida

Division	24-pdr	18-pdr	10-in mortar	8-in mortar	5.5-in howitzer	Total
1	12					12
2	12					12
3	10	4				14
4				2	16	18
5			8			8
Reserve						
Total	34	4	8	2	16	64

Two routes were used from Vila da Ponte, through Pinhel for the guns and through Trancoso for the engineering stores, both routes having been improved enough to take the traffic.[38]

By 24 November, the 2nd, 3rd and 4th divisions of artillery had arrived at Almeida; the mortars did not arrive until 30 November. Dickson said no division took more than six days to make the trip, but also noted that the weather had been particularly good. That was not likely to last. Because of the shortage of food for the oxen near Almeida, they were moved well to the rear, as far as Lamego, Celorico and Viseu. At this point, only the guns and the engineer stores (which had their own carts) had been moved. The ammunition for the guns would have to follow in subsequent trips, the first of which did not leave Vila da Ponte until 29 November. Dickson noted with some exasperation that the Portuguese artillerymen 'did not grease mortars [carriages the] whole way. Axletrees and wheels in consequence much damaged'. These were the carriages that had just been constructed at Vila da Ponte. He reported to Wellington that new carriages would need to be constructed and set his artificers off to build them.[39]

Another of the hidden tasks of operating a siege train now commenced, the 'gauging' of the shot. Not all round shot were exactly the same size, depending on their age (rust) and where they were cast. Troops were being employed collecting round shot around Almeida from previous sieges. Incorrectly sized shot could reduce power and accuracy, and in the extreme case, could block the barrel, causing an explosion. To be safe, all the round shot would be checked to make sure they were within tolerances. Dickson started this process at Almeida in the first days of December, and eventually, about 8,000 shot were recovered.[40]

---

37 Leslie, *Dickson Manuscripts*, vol.3, pp.506–7, 512. Jones, *Sieges*, vol.2, p.360. I have adjusted the number of 5.5-inch (24-pounder) howitzers to 16 as Dickson had previously noted that four had been given to the field artillery brigades.
38 Jones, *Sieges*, vol.1, p.364.
39 Leslie, *Dickson Manuscripts*, vol.3, pp.509–11, 515, 517.
40 Jones, *Sieges*, vol.1, p.364.

By 3 December, only the ammunition and powder for the first division of twelve 24-pounders had arrived (350 rounds per gun). Most of the 1,100 oxen provided by the commissary to move the guns had been returned to enable the continued supply of the army. Three days later, Dickson was asking for the ammunition of the second division of twelve 24-pounders to be brought up next, but it did not arrive until 13 December, 'the bullocks were so knocked up'. On 10 December, Wellington visited and asked Dickson what ammunition had been brought up 'and hinted an idea of attempting something'.[41] At this point, there were only 350 rounds for twelve 24-pounders present. The oxen to bring up the ammunition for the 3rd gun division (ten 24-pdr and four 18-pdr) did not leave Almeida until 14 December. There were only 53 carts in this convoy. Dickson was extremely worried about the condition of the oxen and ordered that they must be rested for several days. They would not be back for two weeks.[42]

A week later, Dickson noted that the powder was to be moved to Almeida by using the 540 mules that usually carried the spare musket ammunition. The powder barrels and made-up cartridges at Vila da Ponte were to be moved first, then about 700 barrels from Peso da Régua. In this letter (22 December) came the first indication that the number of guns to be used in a siege was being reduced. Dickson stated: 'It is not intended to use the 10-inch mortars or 8-inch howitzers and only ten of the 5½ inch howitzers to diminish the necessity of transport as much as possible. For these howitzers must come 3,000 5½ inch shells with fuzes, etc., but I wish all the round shot to come before them.'[43]

It was becoming clear that there was not enough transport to move the full siege train. The loss of the large mortars and howitzers would have a negative impact on the siege. Dickson also commented that 'fascines and gabions are making', it was apparent that something would be happening soon.

The decision on transporting the powder from Peso da Régua changed again on 25 December when Wellington wrote to Dickson confirming that the powder would go up the river 'as far as it can be brought'. The mules would continue bringing the powder from Vila da Ponte. The constant changing of the plan reflected the challenges that were being faced in moving the siege stores forward.[44] Captain George Ross RE had been ordered on 6 November to investigate the river Douro above Peso da Régua, and his report was dated 26 November.[45] Ross concluded that it would be possible to bring boats up to Barca d'Alva from where there was a reasonable road to Ciudad Rodrigo. A great deal of work had been carried out by the port wine companies to clear the river for their boats to transport the wine to Porto. However, this did not mean the journey was easy. The river was still rocky and rapid. Without local knowledge, boats would be damaged or sunk. Even with local knowledge, it

---

41   Leslie, *Dickson Manuscripts*, vol.3, p.515.
42   Leslie, *Dickson Manuscripts*, vol.3, p.518.
43   Leslie, *Dickson Manuscripts*, vol.3, pp.524–29. The shells were not transported when the decision was made not to use the howitzers.
44   Leslie, *Dickson Manuscripts*, vol.3, p.528.
45   TNA: WO55/1561/11, Report upon the Douro from the ferry of St Martinho to the ferry at Barca d'Alva.

was difficult. But it would take two weeks to move the powder 21 leagues (100km) upriver to Barca d'Alva and then by road the 90km to Ciudad Rodrigo. It is not clear if the powder arrived in time for the siege.

On 31 December, Dickson sounded a bit more positive reporting that the 'ammunition now comes in pretty quick. Lord Wellington thinks in about a fortnight we shall have sufficient here to commence operations'. The following day, Dickson knew there was not going to be that much time and there would be less ammunition than hoped.[46]

Apart from the huge tasks of getting the siege train to Ciudad Rodrigo, there were other equally essential tasks that needed to be completed.

Training the Troops
The sieges of Badajoz in 1811 showed Wellington that there were insufficient trained sappers, and he ordered Captain John Burgoyne RE to start training soldiers in this role to ensure some experience was available when Ciudad Rodrigo was attacked. Burgoyne received an order on 28 July to start the training.[47] Later, Burgoyne commented:

> My principal business now is training 200 men of different regiments to the duties required in a siege, which, to our disgrace and misfortune, we have no regular establishment equal to, notwithstanding the repeated experience of the absolute necessity of such a corps to act under the Engineers in a campaign ... The undertaking I am set about will only be temporary and will supply very imperfectly this deficiency.[48]

The training was started six months before the siege and was interrupted a number of times to respond to French operations. John Jones' diary noted that Lieutenant Peter Wright RE had also been ordered up 'to assist in training the men to sap'. Burgoyne did not report starting the training 'in fascine and gabion making and sapping etc' until 15 August 1811. In early September, Burgoyne 'proposed to Headquarters, to pay at six pence apiece for ballast baskets, to contain a cubic foot of earth, to such of my people as choose to make them ... Ten pounds would be well spent ... in the event of a siege'.[49]

This idea was approved, and 400 were made. He later complained that when the French advanced to resupply Ciudad Rodrigo they had taken 'the trouble to burn every stick of my fascines and gabions'.[50] Burgoyne's training was interrupted by this French advance and did not resume until November.[51]

---

46 Leslie, *Dickson Manuscripts*, vol.3, p.534.
47 REM: 4201–68, Burgoyne's Diary 1810–12.
48 George Wrottesley (ed.), *Life and Correspondence of Field Marshal Sir John Burgoyne, Bart.* 1st ed. (London: Richard Bentley, 1873), vol.1, p.137.
49 Wrottesley, *Burgoyne*, vol.1, p.136 and REM: 4601–72, Burgoyne to Squire, Albergaria, 1 September 1811.
50 Wrottesley, *Burgoyne*, vol.1, p.145 and REM: 4601–72, Burgoyne to Squire, Albergaria, 4 October 1811.
51 Wrottesley, *Burgoyne*, vol.1, p.149 and REM: 4201–68, Journal, p.168, 11 Nov 1811.

On 24 August 1811, Fletcher and Jones visited Burgoyne at Albergaria to 'inspect the progress of the party learning to sap'.[52] The next day, Lieutenants Anthony Emmett and William Reid were ordered from Almeida to assist in the training work. They had originally been ordered up to assist in the repairs of Almeida. If three more scarce engineer officers were sent to assist Burgoyne, then this work was judged to be vital by headquarters.

Fletcher visited again on 10 September 1811 and the next day ordered Ross, who was attached to the 1st Division, to start a similar training process with another 200 soldiers. A small number of RMA artificers were sent to assist.[53] Jones' manuscript diary confirms that this training was also ordered in the 4th Division.[54]

These partially trained sappers were very useful during the siege but as Burgoyne said, they were a temporary fix, not a solution. The solution would not arrive for another year.

Bridging the Águeda

To get the guns to Ciudad Rodrigo, two major rivers needed to be crossed, the Côa near Almeida and the Águeda near Ciudad Rodrigo. As part of the early planning for the siege, Wellington had ordered the riverbed of the ford near the broken bridge at Almeida to be improved. In addition, by mid-October, the bridge over the Côa at Almeida had been temporarily repaired, providing two routes across the river.[55] The larger challenge was to get the equipment across the Águeda. The only suitable bridge was at Ciudad Rodrigo and under French control, so an alternative was needed. The Águeda, like many Iberian rivers, was in many places steep-sided, rough and prone to large changes in water level.

On 18 November, orders were issued for 148 tradesmen to be selected from the allied divisions and to report to Major Henry Sturgeon, Royal Staff Corps, at Almeida. Their task was to build a trestle bridge capable of crossing 400 feet (130m). To enable the bridge to resist the winter torrents, each trestle was filled with stone and attached to a pile driven into the riverbed.

The material prepared for the bridge included:

30 trestles
500 running feet (160m) plank, each piece 14 feet (4.5m) long
160 beams, 18 feet (5.8m) long 5 x 10 inches
1 pile-machine on wheels to work in water
180 fathoms (350m) of strong chain.[56]

---

52 REM: 5501-59-2, 24-25 August 1811.
53 REM: 5501-59-2, 10-11 September 1811; REM: 4501-86, Ross to Dalrymple, Nave de Aver, 24 September 1811.
54 REM: 5501-59-2, 17 September 1811.
55 Leslie, *Dickson Manuscripts*, vol.3, p.489.
56 Howard Douglas, *An Essay on the Principles and Construction of Military Bridges and the Passage of Rivers in Military Operations*. 3rd ed. (London: John Murray, 1853), pp.279-80.

The place selected for the bridge was at Molino de Flores near Marialva. The order to lay the bridge was given as part of the general orders on 1 January 1812. Until it was ready, the troops would have to ford the freezing river. The guns would not need it for several more days.[57]

Preparation of Siege Materials

On 17 December 1811, Wellington issued orders to prepare materials needed for the siege. Like all the other aspects of preparing for a siege, this was a significant task. Once prepared, the material would need significant transport to deliver to the front.[58] Troops from the infantry divisions were assigned to prepare them and were paid for their effort. The construction was controlled by engineer officers who monitored the quality of the product.

Table 7: Preparation of siege materials

Item	Number	Weight (kg)	Total weight (kg) estimate
Fascine, 6ft long by 1ft thick	2,500	25	62,500
Gabions 3ft high 2ft 3in diameter	2,000	40	80,000
Tracing fascines 4ft 6in diameter	1,800	8 (est)	14,400
Splinter proof timbers	400	100 (est)	40,000
Sleepers for platforms	200	50 (est)	10,000
Large gabions	30	75	2,250
Fascine pickets	7,000	3	21,000

The total weight of this material would have exceeded 200 metric tonnes and would have required approximately 800 cart trips to move.

The materials were prepared close to where the divisions were bivouacked, and this meant that in the worst cases, they were 40km from Ciudad Rodrigo, a possibly unnecessary increase in the demand to transport.[59]

The Siege

On 28 December, Wellington met with Dickson and Fletcher and determined to 'commence the siege instantly he could get up the smallest possible proportion of stores and ammunition'.[60] The decision to commence the siege appears to have been brought forward when Wellington received intelligence that substantial French troops were being withdrawn, many to support *Maréchal* Suchet's attack on Valencia. Wellington hoped that his rapid advance might draw the French troops back and relieve the pressure on the defenders in Valencia. If this did not work, then Wellington would be able to attack Ciudad Rodrigo whilst the French forces in

---

57 Mark S. Thompson, *Wellington's Favourite Engineer* (Warwick: Helion, 2020), pp.160–161.
58 Based on, Jones, *Sieges*, vol.1, pp.89–90.
59 Thompson, *Wellington's Favourite Engineer*, p.160.
60 REM: 5501–59–2.

the area were reduced.[61] On 1 January 1812, Wellington issued detailed orders for moving the material forward for the siege with the intention of investing Ciudad Rodrigo on 6 January. There were four groups of items to move. At Almeida were the siege guns, powder and ammunition. Also at Almeida was the trestle bridge prepared by Sturgeon. In the area to the west of Ciudad Rodrigo, was all the siege materials prepared by the allied troops. The last was the powder being moved by boat from Peso da Régua.

This unexpected change meant that many things were not ready, the biggest challenge being that the oxen to move the guns were 40km or more away from Almeida. Most of the mortar and howitzer ammunition was still at Vila da Ponte or Lamego due to the transport shortage. There were enough carts available to move the material from Almeida, and Dickson prioritised the movement of the equipment in the order: (1) guns, (2) powder and ammunition for the cannon, (3) ammunition for the howitzers and mortars.[62] This final item was not delivered as Wellington carried out the siege without howitzers and mortars. The consequence was that there were no guns for counter-battery fire or to clear defenders off the walls. This will have led to greater casualties amongst the attackers, both during the siege and the assault.

To make matters worse, it started snowing and the roads became nearly impassable.

The collection point for all the material and equipment was Gallegos, about 15km west of Ciudad Rodrigo. It is difficult to tell what transport was available, but it was in the region of 250 carts and 300 mules. The bad weather continued and, on 5 January, Jones noted that it was taking the carts four days to do a return journey from Almeida to Gallegos. The investment date was put back to 8 January.[63] Dickson noted with some concern on that date that the oxen had still not arrived to move the guns, and he estimated that it would take three days to move them to Ciudad Rodrigo.[64] The oxen arrived the following day and by 11 January were in the artillery park at Ciudad Rodrigo, although only the 38 cannon (thirty-four 24-pounders and four 18-pounders) with two days' supply of ammunition were moved. Dickson noted there was a further two days' supply of ammunition at Gallegos. Carts were assigned to bring forward ammunition from Almeida to Gallegos, and mules then brought it to the siege to keep two days' supply available.[65]

On the night of 13/14 January, the first siege guns were moved into the batteries and the guns opened fire that morning.

This chapter will not deal with the siege. There are many good accounts of the prosecution of this siege, although I would refer you to my accounts in *Wellington's Engineers* and, more recently, *Wellington's Favourite Engineer*. Both of these books take a critical view of the operation using the latest research and much unpublished material.

---

61   Gurwood, *Dispatches*, vol.8, pp.524–525, Wellington to Liverpool, 1 January 1812.
62   Leslie, *Dickson Manuscripts*, vol.3, pp.501–530.
63   REM: 5501-59-2, entry for 8 January 1812. Frustratingly, there is a gap in Dickson's published journal from 8 to 22 January 1812.
64   Leslie, *Dickson Manuscripts*, vol.3, p.566.
65   REM: 5501-59-24, Dickson's account of siege of Ciudad Rodrigo.

Summary

The siege of Ciudad Rodrigo is claimed to be the best-managed allied siege during the Peninsular War. It was planned in great detail several months in advance to allow the huge amount of material to be gathered.

The new siege train was a great benefit, but the logistical challenges of moving the ammunition, powder and stores far outweighed the difficulty of moving the guns. The harsh Iberian terrain, with its inability to feed man or beast when concentrated, increased the challenge of moving the material for the siege.

Wellington took a huge risk starting the siege without the howitzers and mortars, and with limited ammunition due to the transport constraints. It could have gone horribly wrong if the weather stayed bad through the first half of January 1812. Fortunately, the excellent cannon and an ineffective governor contributed to the rapid conclusion of the siege. Success always glosses over the problems that happen during an operation. There is no doubt that allied casualties were higher than necessary due to the lack of counter-battery and anti-personnel fire from mortars and howitzers.

The initial planning for the move of the siege train was excellent, with all the components being in place by November 1811. The later planning and the execution were less effective as the bad weather and the difficulties of keeping the large numbers of animals and civilians healthy and available became increasingly challenging. However, the siege was still a major success for Wellington and gave him more options on how to prosecute the war during the rest of 1812.

A final point. When someone asks why Wellington did not move a siege train from Lisbon to Badajoz in April 1811, when the fortress was unexpectedly taken by the French in March 1811, remember what needed to be done. First, the siege guns had to be available (they were not). Secondly, substantial civilian transport had to be arranged to move the equipment (it was not available). Thirdly, it would take weeks to move the guns and ammunition to where it was needed. This was not something that could be done quickly.

Bibliography

Royal Engineers Museum (REM)
REM 2001-149-11, Report on Almeida.
REM 4501-86, Letters from Ross to Dalrymple.
REM 5501-59-1 Jones journals 1810-12.
REM 5501-59-2, Jones journals 1810-12.
REM 5501-59-24, Dickson's account of siege of Ciudad Rodrigo.

The National Archives (TNA)
WO55/1561/11. Report on river Douro

British Library
BL Add MS41962, Pasley papers

Books
Douglas, Howard. *An Essay on the Principles and Construction of Military Bridges and the Passage of Rivers in Military Operations*. 3rd ed (London: John Murray, 1853)
Gurwood, J.G. (ed.), *The Dispatches of Field Marshal the Duke of Wellington, KG, during His Various Campaigns … from 1799 to 1818*. New Edition (London: John Murray, 1837)
Jones, John. *Journal of the Sieges Carried on by the Army under the Duke of Wellington in Spain During the Years 1811 to 1814*. 3rd ed. (London: John Weale, 1846)
Lipscombe, Nick. *Wellington's Guns. The Untold Story of Wellington and His Artillery in the Peninsula and at Waterloo* (Oxford: Osprey, 2013)
Leslie, John H. (ed.), *The Dickson Manuscripts, Being the Diaries, Letters, Maps, Account Books, with Various Other Papers of the Late Major General Sir Alexander Dickson* (Cambridge: Ken Trotman, 1991)
Summerfield, Stephen. *British Iron Heavy Ordnance for Garrison, Naval and siege (1727-1860)* (Godmanchester: Ken Trotman, 2019)
Thompson, Mark S. *Wellington's Favourite Engineer* (Warwick: Helion, 2020)
Thompson, Mark S. *Wellington's Engineers. Military Engineering in the Peninsular War 1808-14* (Barnsley: Pen & Sword, 2015)
Wrottesley, George (ed.), *Life and Correspondence of Field Marshal Sir John Burgoyne, Bart*. 1st ed. (London: Richard Bennett, 1873)

5

The Siege Lords: British Siege Warfare in India, 1792–1805

Joshua Provan

When making themselves masters of India, the British often looked to the Mughals as a useful example of how to conquer and rule. Indeed, they would eventually promote themselves as the logical successors to the House of Timur in South Asia.[1] In order to do this, they needed to prove themselves worthy of the Mughal legacy.

By the late eighteenth century, the East India Company (EIC) was well on its way to achieving this goal and relied on several principles of military operations that were both peculiar to the European understanding of war in the east, and recognisable to the so-called country powers as acts of superiority reminiscent of the Mughal domination that formed such a prominent place in the history of northern and central India.

This was never clearer than when it came to siege warfare. In principle, the rules that had governed the reduction of fortresses in Europe since the late seventeenth century were applied successfully in India, and with the exception of mud-forts, these answered well. By being able to gather and transport all the paraphernalia of a besieging force to wherever they chose and then slowly and methodically forcing the surrender of the city, fortress or fort, they demonstrated the fiscal, scientific, economic and military power of the Company, and as the EIC expanded another, in some ways more dramatic and dominating method was growing extremely popular with British commanders.

Due to extraordinary examples of audacity and wit such as seen in the escalade of Gwalior during the First Maratha War, by 1780 the British prided themselves on being able not only to vanquish vast hosts of enemies, but on taking any fortification, town or city they chose to take, no matter how numerous the garrison, nor how strong the defences, or how meagre their own means to reduce them. Quite apart from physically showing their power in long formal sieges, where the tableau of a

---

1 This connection was made by many later Victorian writers, but for more on the subject of succeeding to the Mughals, see William Dalrymple, *The Anarchy: The Relentless Rise of the East India Company* (London: Bloomsbury, 2019) and John Keay, *The Honourable Company* (London: Harper Collins, 1991).

state's strength could be displayed, it became increasingly common for the British to assert their dominance by rapid, almost careless, highly aggressive military operations that strove to storm an objective as quickly as possible.

A certain amount of political and military power was derived from the possession of mighty fortresses, and every kingdom tended to have at least one place that was deemed legendarily impregnable.[2] By the late eighteenth century, the British were renowned for their disciplined armies of Sepoys and Europeans, so it was understandable when they defeated opponents in the open. When they were able to overcome defences that were supposedly impossible to reduce, it truly seemed like nothing could stop them.

From a British perspective, the armies ranged against them were often so vast and well mounted compared to their own that decisively defeating them was out of the question. Indian armies could be comprehensively broken, but a single field battle did not guarantee meaningful peace talks, whereas the capture of forts and fortresses often had a greater impact. If no citadel was safe from the British, then there was little point in fighting them. By the turn of the century, their reputation as the greatest fortress breakers since the Mughals was well established, allowing the further development of what might be termed the aggressive Company Style.

By the last quarter of the century, the EIC, growing in capability to wage large-scale wars, came to rely a great deal on the threat of their prowess as masters of siege warfare, and in the years following the fall of Mysore, specifically accelerated siege warfare. This chapter will show this style, and some of the other forms that this prowess took, and its progress into the beginning of the nineteenth century, where it will be seen just how important that reputation had become to upholding the British position in India.

Seringapatam

The third conflict with Mysore was not the same epic struggle as that which ruined the health of Sir Eyre Coote. It was important because it stabilised the company's position in India. Charles, Earl Cornwallis, had brought the Peshwa of the Marathas and the Nizam of Hyderabad into an alliance, forming a combined army to reduce Mysore, for whom there was no real hope of victory. Tipu was not his father, Hyder Ali, and in the end, Britain's favourite foe in India was forced to surrender a large piece of territory to the Company and its allies in 1792. Cornwallis had no interest in destroying Tipu; his campaign had been fraught with difficulty, and it so happened that his vision for India was more in line with maintaining a balance of power that would be upset if Mysore were to become a Company possession, and so Tipu was allowed to negotiate a peace.

The idea of keeping a balance of power in India was dispensed with when Lord Wellesley became Governor General. The new incumbent at Fort William wished to

---

2   Gwalior, Srirangapatna, Gawilghur and Bharatpur were all described as being thought to be impregnable by their defenders, to that list could be added any fortress protecting a capital, and usually, any capital.

end the complicated business of keeping everyone in equilibrium by making them all dependencies of the EIC. 'The balance of power in India no longer exists upon the same footing on which it was placed by the peace of Seringapatam [1792],'[3] and so under the auspices of driving out the last French diplomatic and military enclaves in South Asia, Wellesley began to create a Company *pax*.

This essentially meant a significant loss of sovereignty to any kingdom that accepted what were termed subsidiary alliances, and this, along with Tipu's evident leanings towards France and any other enemy of the British, made another war with Mysore inevitable after Wellesley came into office. When war broke out, the British went to great lengths in order to ensure the fall of Mysore and, in doing so, made a great statement. Seringapatam was the stage on which would be enacted Lord Wellesley's forward policy, and like any good showman, he lavished a great deal of finance and supplies on the production, which included a large contingent from Hyderabad and the theoretical cooperation of the Marathas. The South Asian world and indeed that of Europe was encouraged to sit up and watch as the Company went to war, and laid siege to the island fortress of the Tiger of Mysore.

Compared with the other operations covered in this chapter, the great siege of Seringapatam in 1799 was an almost refined affair. This did not mean it was a simple matter of appearing before the fortress and letting the science of engineering and siege craft take their course, however.

All around devastation prevailed, since Tipu had burned the countryside, which had not fully recovered from the 1792 siege, putting strain on the long supply lines which would have to hold out until the rains came or the city fell. The city itself was strong, sitting on an island in the river Cauvrey, with an imposing fortress at its western side and the walled city to the east. After a slow start from both defenders and besiegers, Lieutenant General Harris received reinforcements in the form of the Bombay Army, which would operate on the northern bank, while the Madras Army would do the same from the south, each targeting the fortress on the west side of the island from opposing sides.

Although Harris was behind schedule and racing the monsoon, Tipu did not even try to harass the British until after the Bombay contingent arrived, by which time Harris had mostly cleared the southern bank and broke ground on 20 April. Six days later, the British having successfully beaten back some serious sorties, Colonels Wellesley and Sherbrooke secured the last Mysori entrenchments, and a deadly seesaw battle developed over the final position that commanded one of the bridges over the river. Between 26 and 27 April, the British suffered 300 casualties in achieving this objective.[4]

Over the next two nights, the breaching battery was established and the heavy guns were emplaced. When it came to breaking down the wall, a crafty fire plan was devised for maximum speed and impact. Several more enfilading batteries and a false bombardment, targeting a different section of wall, had the effect of distracting Tipu from the actual point of attack.

---

3   Montgomery Martin, *The Despatches, Minutes and Correspondence of the Marquess of Wellesley, K.G. During His Administration in India* (London: John Murray, 1836), vol.I, p.28.
4   Rory Muir, *Wellington: The Path to Victory* (New Haven: Yale University Press, 2013), p.84.

The legitimate breaching batteries opened fire on 2 May, and achieved a practicable breach within 24 hours. After a detailed reconnaissance of the breach and the fords over the river, the assault was ordered for 1:00 p.m. on 4 May, so as to once more wrong-foot the defenders. It does not appear that Tipu was given the option to surrender, and it is regrettably the case that the ordinary rules of siege warfare, which required a summons to surrender upon completion of a practicable breach, were rarely observed in India, as will be seen from the examples given below. Generally, the defenders were required to capitulate at the start of a siege, or without summons once the walls were down, but it is unlikely Tipu would have gone quietly even if Harris had offered to spare his city. Major General Baird led the assault, and the Sepoys and Europeans mounted the breach in fine style.[5] Tipu died fighting, famously preferring to live one day as a tiger than a lifetime as a sheep.[6]

What was less 'stylish' were the horrors that followed, and were likely far worse even than the sacks of Ciudad Rodrigo and Badajoz, although they served a purpose to the British. In the words of Lieutenant Patrick Brown, the city and fortress

> … presented a shocking sight scattered all over with dead black bodies cut and mangled in a dreadful manner and here and there lay an unfortunate European … We stayed in the fort all night under arms where there was a terrible state of confusion, several of the houses on fire, and the soldiers running about through the streets with drawn sabres and lighted torches, dressed up in silk clothes etc which they had plundered and compelled the poor inhabitants to show them where their treasure was concealed. Some officers and many private Europeans made their fortunes that night …[7]

William Harness, who had no sympathy for Tipu, arrived to take command of his regiment almost two weeks later and recorded that sales of the loot were still going on,

> In walking round the fort my amazement is at every step heightened at the possibility of a Fortress of such strength, defended by half an empire, falling before the courage and conduct of a handful of determined Men. The carnage of a sacked capital may be imagined. Thousands threw themselves over the rampart into the Cauvery. It's rapid stream is infected for miles. Col. Wellesley is put into the place; while calling on him the day after I arrived, the provost reported to him that he had buried nine-thousand and the works are far from being cleared. It appears that the attack was made at a most critical moment, that all the provisions were exhausted and that the troops that led the storm went absolutely fasting. The town was

---

5   Harris lost 1,531 casualties during the siege, slightly more than half were Europeans, as he and Lord Lake tended to depend most on them for storming duties.
6   For more on the siege, readers are advised to seek out Rory Muir, *Wellington: The Path to Victory* and M. R. Howard, *Wellington and the British Army's Campaigns in India, 1798–1805* (Barnsley: Pen & Sword, 2020)
7   Muir, *Wellington*, p.85

richly stored. The money and jewels that are now sealed by the Prize Agents exceeds a million Sterling.[8]

News of the sack and the tremendous treasure, plus exhibits of it, spread through southern India within the week, first through official channels and then into the presidency newspapers. Lady Henrietta Clive wrote to her brother, the 2nd Earl of Powis, in a letter written sometime after 12 May that:

> The plunder of Seringapatam is immense. General Harris will get between £150,000 and £200,000. Two of the privates of the 74th have got £10,000 in jewels and money. The riches are quite extraordinary. Lord Clive has got a very beautiful blunderbuss that was Tipu's and much at Seringapatam. Some of the soldiers have got 20,000 pagodas; some have ten thousand pagodas, and one a large box of pearls. I should like to have the picking of some of the boxes. There was a throne of gold, which I am very sorry to say they are breaking to pieces and selling by parts. Lord Mornington has presented me with one of the jewelled tygers [sic] from the throne.[9]

Lieutenant Brown was nothing short of prescient when he wrote, 'I don't think there will ever be such a prize taken in India again.'[10] He was correct in more ways than one, as a siege on this scale would not be seen for another 28 years, although in 1805, the British would besiege a city of equal size, but we are getting ahead of ourselves.

When the ageing Maratha statesman, Nana Fadnavis, whose second sight might sometimes be mistaken for hindsight, heard of the fall of Mysore, the scale of what had happened was not lost on him. It seemed to dawn on the 'Maratha Machiavelli' that what had really occurred was not simply the elimination of a common enemy, but the beginning of the end for the sovereign powers of India. Nana warned Bajirao II Peshwa, 'Tipu is finished; the British power has increased; the whole of east India is already theirs; Poona [sic] will now be the next victim. Evil days seem to be ahead. There seems to be no escape from destiny'.[11]

Sasni and the Ceded Provinces

There was an old saying in India that the Maratha had his spear, the Afghan had his sword, the Sikh had his matchlock, and the Englishman had his cannon.[12] The implication was that each felt he was invincible with his favourite weapon. The siege

---

8   Caroline M. Duncan-Jones (ed.), *Trusty and Well Beloved: The Letters Home of William Harness, an Officer of George the Third* (London: S.P.C.K, 1957), pp.133–134.
9   Henrietta Clive, Nancy K. Shields (ed.) *Birds of Passage: Henrietta Clive's Travels in South India 1798–1801* (London: Eland, 2016), p.82
10 Muir, *Wellington*, p.85.
11 Govind Sakharam Sardesai, *New History of the Marathas* (Bombay: Pheonix Publications, 1946), vol.III, p.354.
12 W.H. Allen, *An Illustrated Handbook of Indian Arms* (London: William H. Allen & Co, 1880), p.127.

of Seringapatam showed that once the British had opened a battery, they were indeed formidable with their cannons, and that Tipu was courageous to elect to defend the breach rather than ask for terms. The fall of Mysore was a pivotal moment for the EIC, and was recognised across India as another mark of the ambition and power of the Company. Many, however, chose Tipu's way.

In 1801, the Newab of Oudh ceded several provinces to the EIC in return for military protection. The threat of the Durrani Empire of Afghanistan had been potent for most of the 1790s, with fears that Zemaun Shah would launch a full-scale invasion of Hindustan. According to the British, in 1802, these provinces were controlled by a number of *Jemadars* (local landowners) who rejected the arrangement and were refractory in their conduct, who further committed 'sad depredations ... setting the laws of government at defiance.'[13]

Chief among them was the Rajah of Sasni (Sarsney), who controlled a number of forts and a sizeable army.

On 28 August, at least one company of grenadiers from the 1st Battalion, 2nd Bengal Native Infantry, under Lieutenant Pester, Ensign Marsden and an unnamed Subadar, were dispatched 32 miles from Shikohabad to stamp out a rebel stronghold.[14] The troops marched for 12 hours, arriving at a village called Assnayder at 4:00 a.m. the next day. Here, Pester was told that his quarry was 500 strong, and two miles away at a walled village called Camney, and was aware of his advance. The lieutenant rested his troops for half an hour, but then as dawn became a suggestion in the east, he received a taunt from the rebels offering him safe passage if he retreated, but no quarter if he continued into their territory, adding that they would make the fine red coats of the Sepoys into shoes once they were all dead, as an extra insult.

Pester read the insolent defiance aloud to his grenadiers, who decided to punish the low born rabble who dared to mock them in such a manner, promising to convince the rebels that they had made a grave mistake. Satisfied he had encouraged the proper fighting spirit, Pester pressed onwards, arriving in sight of the place at 5:00 a.m., where he rode ahead with a guide and was saluted with the flash and crackle of musketry from the walls as he investigated the defences.

Riding back to the column through the pelting lead, the young officer, on his first independent command, evaluated the situation. As was common in Hindustan and the Deccan, small villages were defended by stout mud walls. In some of the larger cases, these were set with towers to ward off raiders. Camney was not large, but it did have a wall and an inner tower, accessed by two gates and defended by a sizeable force of undetermined origin, whom Pester seemed to think were little better than brigands.[15]

Pester regretted the lack of a small field-piece with which to blow the gates, but not wanting to be found lacking in dash in his first command, and with no weapon larger than an infantry musket, nor a tool stronger than the butt of the same, and the shoulders of his grenadiers, he decided he would assault the gates immediately.

---

13 John Pester, *War and Sport in India* (London: Heath, Cranton & Ouseley, 1913), p.13.
14 According to Pester the 2nd BNI had two grenadier companies.
15 Pester, *War and Sport*, pp.13–15.

Lieutenant Pester separated his men into two divisions, leading one himself while Marsden led the other.[16] They converged on opposing gates with a commendable show of timing, though incurring casualties from the heavy fire directed from the walls. Once up against the defences musket butt and shoulders smashed open the gates, whereon Pester was unable or disinclined to dissuade his men from shooting and bayoneting the garrison practically to the last man; he calculated later that 200 had fallen in the battle. The last remaining stronghold was a round tower that commanded most of the village. Gathering his Subedar and 30 men, the lieutenant hurried to blockade it. Four men dropped to enemy fire behind him as they ran, but here he linked up with Marsden, whose coat was torn up by musket balls, and two men were ordered to procure pickaxes and shovels from the surrounding villages. In the end, when the rebels overheard the plan to dig them out, they immediately surrendered.

Pester's blood was running high, having lost a good many men (though he did not specify a number), and he briefly considered making an example of the enemy. Exterminating the garrison would not have been an unusual act. However, the rebels having thrown their muskets out of the loopholes, the lieutenant could not bring himself to slaughter them. Gaining entrance, they bound the remnants hand and foot, dragged them clear, and burned the village to the ground before setting out to return to Assnayder. He had not gone far when he was informed by a local inhabitant that the surrounding villages had taken up arms and intended to rescue the prisoners or die in the attempt. Pester always took great care to inspect musket locks and cartridge boxes, and found that most of his men had only two or three rounds left.

Due to the dire situation, the sergeant in charge of the prisoners was ordered to kill his wards immediately if they should be attacked. Pester then selected an advantageous position from which he intended to charge the first enemy to show his face. As it happened, no hostile force appeared, and the grenadiers made it back to Assnayder where, 'out of dread of us,' the locals provided a nourishing meal.[17] Pester's detachment returned to Shikohabad on 30 August unmolested, having been on their feet, and in the same clothes, on the march and in action, mostly without sleep, since the 28th. Poor Ensign Marsden never recovered from the fatigue, and perhaps contracting a fever in his weakened state, died soon afterwards.

This was the opening of operations in the so-called Ceded provinces, which by December had taken on a severity that warranted the deployment of a sizeable contingent, which included the 1st and 2nd Battalions of the 2nd BNI, unspecified battalions of the 8th and 12th BNI, the 3rd Bengal Native Cavalry, and a small siege train which included two howitzers, all operating under the 1st/2nd BNI's commander, Lieutenant Colonel Blair.[18]

The fort belonging to the Rajah of Sasni was made the target of this force, as he had 'refused to accede to the established rulers of our government, and had determined on disputing the point with us, and neither to surrender his forts or disband his troops.'[19]

---

16 Pester, *War and Sport*, p.15.
17 Pester, *War and Sport*, p.17.
18 *Pester, War & Sport*, pp.24–25.
19 *Pester, War & Sport*, pp.24–25.

The Bengal troops marched for nine days, at times violating Maratha territory, before coming in sight of the Rajah's fort at Bidgie Ghur, which was bypassed in favour of Sasni, which itself stood five miles distant. This was a more serious affair than the bandit stronghold of Camney. 'The walls we perceived were covered with fighting men, and with our glasses we could count a great many heavy guns mounted on the ramparts. A large body of cavalry and infantry were also encamped on the glacis of the fort, which appeared to be strong and very lofty.'[20]

There followed about 13 days of negotiations, which ended on 23 December, when Pester and some friends breakfasted on cold beef and then 'went down to be shot at'. The British beat the Rajah's troops out of some villages and groves fronting the fort in order to occupy the ground necessary to begin their trenches.[21] The sepoys and their pioneers toiled through the night so that by morning, all was completed and the attackers could remain in cover. Seeing this, the Rajah or his commander ordered a sally, and about 200-500 men with close combat weapons left the fortifications and took up a position in a field of grain or corn, where they waited for midday. At noon, the fort's artillery opened a very heavy fire on the enemy entrenchments, taking some of the British officers by surprise. At this signal, the whole of the sallying party rose up and threw themselves forward, yelling furiously, towards the most advanced work, which was occupied by Pester and his grenadiers.

The lieutenant remained cool, even though the officer in charge of the reserve lost his composure and was found skulking in a rear trench, unable to command his men. According to Pester, however, he was able to command all the troops within the vicinity to reserve their fire until the last possible moment when a stunning discharge of musketry caused the impetus of the rush to be lost, and after a brief struggle, the Rajah's men were repulsed.

By 1 January, three batteries had been created, and no more attacks had been received. However, the detachment lost men every day to enemy fire. The batteries opened on the 4th, but though they consisted of a grand battery, an enfilading battery and a further battery for the suppression of the ramparts,[22] the walls 'were exceedingly well cemented and came down very sparingly.'[23]

Due to the strength of the defences, the long range of the grand battery, and the activity of the Rajah's artillerymen, the bombardment dragged on without much result for three days. Pester noted that the 12- and 6-pounder battery, located at a grove of trees, had taken such heavy fire that the shady flora that had once typified the position had been shredded to pieces or knocked down. The Rajah's gunners were extremely adept at their work, having at their disposal at least 40 guns, ranging from 6-pounders to a monstrous 80-pounder that the British sepoys called 'The Lightning.' They quickly found the range and elevation to strike the British embrasures, prompting some of them to be blocked up with fascines, with tragic results.

On 7 December, a sergeant major of the Royal Artillery was cut in half by a shot that went through an embrasure while taking a rest by sitting on the trails of a

---

20   Pester, *War & Sport*, p.27.
21   Pester, *War & Sport*, p.27.
22   18-, 12- and 6- pounders with howitzer support.
23   Pester, *War & Sport*, p.31.

gun carriage. Pester was witness to another tragic incident in the enfilading battery, where a personal duel of cannons was taking place. The defenders had brought up and sited a new piece, which greatly annoyed the British, and much effort was put into knocking it out of action. A lieutenant was in charge of the battery and was standing on the parapet to better observe the fall of the counter-fire, when the one of the lookouts called out 'Shot!'

The gun had already killed a number of the besiegers, and its operators were not to be trifled with. The lieutenant at once leapt from the parapet, but as he came to earth, he fatefully passed one of the blocked-up embrasures at the instant the ball broke through the fascines, which took away both his thighs as it continued its path. The horrified onlookers rushed to evacuate the casualty, who died soon after receiving medical attention.

Rain, the occasional sally and roving bands of enemy cavalry operating on the British supply line to Secundera (Sikandra) made progress slow. However, a redoubt was added a little in advance of the first parallel, three days after the lieutenant was killed, and between 12 and 14 January, a breach was created which was declared practicable. With a storm now almost inevitable, the British switched their fire to the defences on either side of the breach, and then struggled to suppress the enemy artillery, which was being dragged into position to defend it. The storm was set for pre-dawn on the 15th and was to consist of the grenadier companies of the entire detachment.[24] The grenadiers were to be led by 40 men with scaling ladders, and all of the storming party were to advance once a false attack had drawn the enemy's attention to the north gate. The grenadiers were ordered to remove their cummerbunds so as to be able to move more freely, and after some debate on whether to remove the flints from their muskets, they primed and loaded. With the coming of dawn, the stormers were in position, and the attack went ahead with great spirit from officers and men. However, the enemy discovered their advance when they were still 40 yards from the glacis, and the ladders were found to be too short. After a brutal struggle at the main breach, the sepoys were ordered to retreat, having taken significant losses as the attack lost momentum.

Disheartening as this was, little importance was given to it considering the strength of the objective and its garrison. Indeed, the reverse ensured the British would try again.

Lord Lake, commander-in-chief in India, arrived on 28 January, accompanied by His Majesty's (HM) 27th Light Dragoons and two regiments of unspecified Native Cavalry – HM's 76th Foot would arrive soon afterwards. Though one might expect this appearance to place a new face on the siege, Lake remained in observation, leading a costly reconnaissance in force with six companies of grenadiers the next day. Lake and Pester were both dismounted by the heavy fire that met them, and it was a bruising introduction to Lake's famous intrepidity. The commander-in-chief did, however, determine that the fort was too well defended to be taken by storm

---

24 In all eight companies as, according to Pester, the Bengal Army at this time still operated with a right and left grenadier company.

without more useless loss, and so planned to take the Pettah instead.[25] The British artillery was therefore directed to remove any impediments to accomplishing this object. On a foggy 4 February, Major David Ochterlorny successfully escaladed the town defences, which was no sooner achieved than reinforcements hurried to the spot to repulse an anticipated counterattack. This reverse had the effect of utterly demoralising the garrison, who took the opportunity to evacuate the fort in the early hours of 6 February, rather than face another British storm.

Bidgie Ghur was then invested by Lake and bombarded between 8 and 14 February, where again the garrison evacuated the fortifications rather than risk being slaughtered in the inevitable assault.

Lake sent some of his troops back to Secundra after this, and pushed on himself with a small force to reduce another fort at Kachoura, which was supposedly prepared to surrender. This information proved to be false, however, as the Rajah of Sasni's son was in command, and the British force allocated to take over was nearly trapped and wiped out after entering the fort. The Secundera force, therefore, had to be recalled before they reached their destination. Lake determined to punish the garrison, which, after seeing their plot fail and coming under bombardment, lost the will to fight and chose to evacuate the fort in the early hours of 21 February. However, the British were prepared, and most of the defenders were cut down by the cavalry in the open, including the Prince.

Pester's account of assaulting the mud-forts is instructive. That something is occurring amidst the company officers beyond inexperience and youthful exuberance can be seen by the actions of his commanding officer, and indeed of Lord Lake. Undoubtedly, this offensive spirit had its origins in self-preservation against larger but more disorganised Indian forces, before whom those with even limited service in South Asia believed it to be disastrous to retreat, and the success that audacity bred was slowly taking on a life of its own.[26] In the case of Pester at Camney, it is striking that an officer of this relatively low rank would elect to sacrifice his men in an escalade without so much as a foot-stool to aid his entrance, rather than delay to obtain the proper apparatus and risk the stigma of being seen as hesitant in the face of an Indian enemy, (let alone suspected brigands). This speaks to the attitude expected of any EIC officer in the field at the dawn of the nineteenth century.

Ahmednager and Gawlighur

Lake's war with the intransigent *Jemadars* was hardly as neat as the siege of Seringapatam, indeed, the assault on Sasni was almost disastrous, but as was fairly typical, the British were able to absorb the reverse and press ahead. The lesson of mud walls and a determined garrison was, however, not learned by the

---

25   Pettahs being the fortified civil quarters of Indian town and villages they were quite often easier to assault.
26   John Gurwood (ed.), *The Dispatches of Field Marshal the Duke of Wellington* (London: Murray, 1840), vol.II, pp.31–32, Wellesley to Close, 23 June 1803.

commander-in-chief, and many might well have put the unsuccessful storm down to a poor selection of target, since the attack on the Pettah proved decisive.

This tactical flexibility in the face of reverses can be seen employed against the Marathas a year and a half later. Ahmednager is a city in western India that was once an important state of its own, but by 1803 it formed part of the Maratha confederacy and was one of the gateways that any army had to pass from Pune to reach the Deccan plateau. At the time of the Second Maratha War, it consisted of a citadel and an extensive town wall. Major General Arthur Wellesley had been preparing to capture it for at least a month before beginning his campaign to, so to speak, enforce his brother's politics by other means.[27] Captain James Welsh of the Madras Army described the place like this:

> The Pettah of Ahmednugger [sic] is a very large and regular native town, surrounded by a wall of stone and mud, about eighteen feet high, and very neatly built, with small bastions at every hundred yards, but no rampart to the curtains; the wall being rounded off at the top, and scarcely broad enough for a man to stand upon. It has several gateways, and many high buildings in the interior, with narrow streets, and mud walls of different compounds, all contributing to aid a powerful defence; but, alas for it's security, the determined spirit was wanting.[28]

The British arrived before these defences on 8 August 1803 and summoned the *Quilledar* (Khilledar) or garrison commander to surrender. This nicety being rejected, Wellesley made his dispositions. With ill health stalking the British camp after prolonged inactivity, and a need to obtain the plentiful stores in Ahmednager for the coming campaign, an assault with scaling ladders went forward. This went badly at first due to a misunderstanding about the nature of the walls, since there was no walk or rampart on the stretches of curtain wall between the towers for the men to gain a footing on after clambering over, as Welsh described. Thus, the first attempt was essentially repulsed. However, despite the initial problem of identifying what parts of the wall to attack, the defences of the town were overcome by the audacity of a few officers, very much of Pester's mould, who gained entrance through the towers. After a short period of disorder where the troops attempted to ransack the town, resulting in the hanging of two sepoys for looting, the British then moved to invest the citadel. A small redoubt was constructed, and the fire of four 12-pounders and two howitzers was enough to convince the Maratha commander that he had no choice but to surrender on 12 August.

Wellesley had little practical experience of leading a formal siege, despite serving as one of Harris' senior subordinates at Seringapatam. His role there had been in clearing out towns and fortifications of enemy troops so the works could be started; as such, the quick resolution to the assault of Ahmednager suited him immensely.

---

27 Gurwood, *Dispatches*, vol.II, p.32, Wellesley to Stevenson 23 June 1803; p.36, Wellesley to Stevenson, 24 June 1803.
28 Joshua Provan, *Bullocks Grain and Good Madeira: The Maratha and Jat Campaigns, 1803–1806* (Warwick: Helion, 2021), p.50.

Things were not so simple when, at the end of his campaign in the Deccan, he encountered the fortress of Gawilghur. Being an efficient planner, Wellesley had been considering the capture of Gawilghur since late August, or early September 1803. This place belonged to the Maratha Rajah of Berrar, who was the last of the principal Maratha leaders to remain in the field after Assaye (so long as one does not count Yashwantrao Holkar who was at this time semi neutral) and was far stronger than Ahmednager due to its advantageous position. It was a true hill-fort sitting as it did on the pedestal-like promontory of a range of mountains. Despite this, Wellesley did not think it any stronger than many others of its type in India. EIC engineer officer John Blakiston described it with a professional eye:

> It consists of a lofty mountain, the plan of which is somewhat in the shape of the figure 8; the smaller end being connected with the table-land to the northward by a narrow isthmus, and the larger circle jutting out into the plain, having the sides separated from the contiguous mountains by deep chasms or valleys. It is thus naturally divided into two forts, that to the southward being the inner one, or citadel, having its sides very precipitous, and only to be approached from the plains; while that to the northward, which is joined to the table-land, forms the outer fort; thus producing a double line of defence on the only weak points. Excepting across the isthmus above-mentioned, and at the separation of the two forts, the walls are not particularly lofty; neither is it necessary that they should be, the scarped nature of the mountain requiring but little additional defence. But, at those weak points already alluded to, the walls were both strong and high, and well flanked with towers, but without having any ditches of consequence.[29]

It also had the curiosity of a double wall that was not at first discernible until the battering began, and similarly, the two parts of the fort were separate from each other and connected by a single bridge of rock that led to another gate. Wellesley had no precise information about any of this when he began to invest it on 3 December 1803, although his engineer officers had given accurate topographical information about where to attack from. Initially, he had to push a division into the mostly trackless mountains to haul the guns into place. It was a task of considerable effort, taking from the 7th to the 12th of December to get the men into position, and one that Wellesley called the most laborious and brilliant feat of the campaign. Although the defenders kept up a harassing fire and attempted some sorties, the British were in complete control of the position. For this operation the British could call on two 18-pounder siege guns in addition to the other artillery, most of which was posted enviably close to the walls, and opened fire on the 13th.

On 14 December, a breach was declared practicable and was carried the next day in under 10 minutes. The double wall posed a problem, but once the attack was in progress, the battalion officers were loath to tamely give up.

---

29 Provan, *Bullocks, Grain etc*, p.83.

Following the fleeing defenders into their fortifications, a chase ensued past the second wall and into the so-called outer fort, where close-range musketry and bayonet work soon saw the streets so choked with dead and wounded in places that it was difficult to open gateways to progress further. With resistance crumbling, the attackers pressed on to seize the outer fort, only to find that it was connected to the inner one via the rock bridge over which ran a path leading to a gate of yet another double wall. Though the engineers had expected this, for the infantry it was something of a shock. Problematically, the defenders refused to surrender, despite being totally cut off now, forcing the British to contemplate either pressing on at once and trying to find a way to continue or bringing up the guns and creating another breach. A few haphazard attempts were unsuccessful before two battalions of Native Infantry took up position and swept the facing walls with file fire. Using this cover, Captain Campbell of the 94th Foot scaled up a ladder and made an entry with his light company, which captured the gatehouse, allowing the rest of the attackers inside, where the slaughter continued until at last the final elements of the garrison surrendered.

Here once again, initial difficulties were overcome by the wit and courage of field and junior officers leading the attack. In the words of Wellesley's ally, Bopa Ghokale, who witnessed the precipitous assault on Ahmednager, 'They [the British] came here this Morning, looked at the pettah wall, walked over it, killed the garrison and returned to breakfast. What can withstand them?' This was certainly the response that the British hoped to inspire by such feats of arms. However, as we go on, it will be useful to observe that though these two storms were audacious, each was the product of careful preparation and planning. Wherever possible, overwhelming firepower was brought against a logical point at relatively close range, and in the case of Gawilghur, was breached within 24 hours. Comparing these operations to Blair's and Lake's siege of Sasni, it is fair to say that, as much as the British might despise their enemy and act out of reputation as much as practicality, a lot still depended on the commander.

Aligarh, Deeg and Bharatpur

When the Hindustan campaign of the Second Maratha War opened, Aligarh was the first target of Lord Lake's Grand Army. The opposition here was light, as the EIC's coffers had been opened and most of the Europeans in the Maratha forces had defected, either out of greed or due to the ultimatum issued by Lord Wellesley threatening dire consequences for any European found in arms against the Company. Here, the loyal garrison had arrested their turncoat commander, and finding the new *Quilledar* obstinate in his intention to defend his post, Lake determined to take it by force. Aligarh was a strong place, with multiple tiers of defences and an enormous wet ditch that the British could not find a way past. Instead, efforts were directed towards the gate, where naturally the ditch could not continue to be a hazard. Much as in other cases with Lake's sieges, the plan to assault the gate did not survive contact with the enemy, and indeed, the attacking force, which went forward with ladders, was almost wiped out as they struggled to get over the walls.

Only when some artillery pieces were brought forward, which blew the gates open, did things change for the British, who were then able to pour inside and neutralise the garrison in what was effectively a three-hour sack. The losses were considerable.

This fortress was taken inch-by-inch according to Lake, but was an impressive feat, with Major General Wellesley marvelling at anyone being able to successfully batter down a gate; something he had never been able to do. In flattering Lord Lake on the day of the storm, the famous cavalryman, James Skinner opined that no fort in Hindustan could stand before the Europeans.[30]

Apart from their medieval grandeur and often inaccessible positions, Ahmednager, Gawilghur, Aligarh and even Seringapatam presented no overly alien aspect to a British general. They were large, old and imposing; conventionally defended by tall walls, bastions and towers, predominantly made of stone and brick, augmented by earthworks and, in some cases, by natural stone. But as the storm of Aligarh, and Seringapatam showed, a fortress in India was only as good as it's ditch, in the former case, Lake elected to avoid it almost entirely, and in the latter, the river which doubled as a ditch was so shallow at the time of the assault as to be nullified as an obstacle. Why Lake did not follow his own example when he came before Bharatpur in 1805 is difficult to say.

Some might have been troubled by the bruising experience of Lake's punitive expeditions against fortifications made predominantly from mud brick. Certainly, some years after the Second Maratha War, one officer of the Bengal Army had identified this type of fortification as so troublesome as to warrant the penning of a memorandum specifically dealing with the reduction of mud forts.[31] However, in the period of expansion under Lord Wellesley, the very name of this class of fortification lent itself to creating the impression of something rudimentary and inferior. 'The Bengal Army, justly appreciating its own pre-eminence, has been accustomed to look with contempt upon troops so far their inferiors as those employed by the native powers, and various instances of extraordinary success has induced them equally to despise their fortifications.'[32]

Pester certainly thought it beneath him as a Presidency officer to so much as await a galloper gun with which to assault Camney, and Blair's contemptuous first approach to Sasni was typical of how the British dealt with mud forts. Indeed, it was likewise the attitude taken towards even the monumental mud walls of the Jat capital of Bharatpur.

The Rajah in question had broken his treaty with the British and, perhaps, goaded by his son, thrown in with Yashwantrao Holkar after the defeat of Daulatrao Sindhia and the principal Maratha Confederates. At the time, Holkar had scored a major victory over a large company detachment and had come close to recapturing Delhi. Many of the smaller states felt that they had misjudged the Marathas and should join Holkar. Lake felt personally humiliated by Bharatpur's defection and made the capture of his fortresses a major priority after the relief of Delhi and the scattering

---

30   James Ballie Fraser, *Military Memoir of Lt-Col James Skinner, CB* (London: Smith Elder & Co, 1851), vol.I, p.267.
31   Anon., *Observations on The Attack of Mud Forts* (Calcutta: Pereira, 1813)
32   Anon., *Observations on The Attack of Mud Forts*, p.2.

of Holkar's forces. Even if the great Maratha general had surrendered, Lake would have chastised the Rajah.

Now isolated, the Jats seemed easy prey. The frontier fortress of Deeg had fallen to assault, though not without considerable loss, but all seemed to be following the usual pattern. An outwork called the Shaburj, which was unprotected by a ditch, had been selected for the attack, and a party of 80 Europeans and 200 Sepoys were sent to begin the entrenchments. The journal kept by the author of the treatise on mud forts ran:

> It was half past nine [on 20 December] when the engineer began to mark out the battery and about half an hour afterwards the parallel; we soon entrenched ourselves; and the battery was completed by daybreak, when the guns opened with great effect, and very soon demolished the small gateway. By sunset we had fired upwards of 500 rounds, which had more effect than the Grand Battery, which opened on the morning of the 17th. We had [worked] nearly three hours before we were discovered; and when (as we supposed) they did discover us, they fired only a few shot in the night. At day break however, they opened a number of guns upon us, but, then, we were secure, and suffered no injury, not a man was touched. December 22, this battery continued to play all this day and by evening the breach was practicable.

The infantry was then able to take the breach with the bayonet between the night and dawn of 23–24 December 1804.[33]

Lake next moved against the city of Bharatpur, rejecting overtures from the Rajah, Runjeet Singh (not to be confused with the Sikh maharajah) and was in position before the walls by the end of the first week of January 1805. Lake wrote that the city and fortress were of

> ... great extent, every where strongly fortified. A mud wall of great height and thickness, and a very wide and deep ditch, everywhere surround it. The Fort ... is of a square figure. One side of that square overlooks the country; the remaining three sides are within the Town. It occupies a situation that appears more elevated than the Town; and its walls are said to be higher, and its ditch of greater width and deepness. The circumference of the Town and Fort is upwards of eight miles, and their walls in all that extent are flanked with bastions at short distances, on which are mounted a very numerous artillery. This place derives a considerable degree of its strength from the great quantity of water which its situation enables it to command. Its ditch being filled with this, presents an obstacle very difficult to be overcome.[34]

---

33   Anon., *Observation on the Attack of Mud Forts*, p.16.
34   J.N. Creighton, *Narrative of the Capture of Bhurtpore in the province of Agra, upper Hindustan, by the forces under the command of Lord Combermore, in the latter end of 1825, and beginning of 1826* (London: Privately Printed, 1830), p.IX.

Most mud forts had presented difficulties to the British, due to the peculiar makeup of the walls themselves and how they broke, and their often large ditches. The hauteur of the attackers in approaching them encouraged either extreme or lax versions of the *attaque brusque*, which was defined by the engineer, Sir John Fox Burgoyne:

> An irregular or accelerated attack (*attaque brusquée*) is one in which the tedious forms prescribed for the reduction of fortresses are wholly or in part dispensed with, and much judgement is required in the general and the engineer to know when it may be applied with effect; that is, to reject each form in precise proportion of the place or the force of circumstances, and no further; for their must be more or less risk of failure in operations so conducted, if applied in excess; whereas nothing ought to be more certain than the result of those that are conducted on regular siege principles.[35]

The Company style of turning a siege around in under a week had yet to be tested against a place as strong as Bharatpur, but if it was to be taken brusquely, (and the reader will have noted that the majority of the sieges mentioned here could be defined in the terms outlined by Burgoyne) the great ditch would need to be overcome, which usually meant directly attacking a gateway. However, for reasons that are not altogether clear given his success at Aligarh, and which would instead seem inspired by Sasni, Lake declined this method and instead chose to begin operations against the Pettah wall. One of his horse artillery officers, Captain-Lieutenant Young, had been dubious about this from the start.

> The circumference of the place was so vast that to enfilade the curtains (the first grand rule in besieging) was impossible, for the prolongation lines would have carried the enfilading batteries a mile or two away to the right or left. The curtains were short and the bastions which projected very far were united to the place, by long and comparatively low necks and thus the high bastion, formed an impenetrable traverse for two or three guns on the neck, which could only be dismounted by chance mortar-shells and where thus reserved to pour deadly volleys of grape, at twenty yards distances, into the ditch and the breaches.[36]

The siege of Bharatpur did not disappoint the expectations of the troops, who looked with dread at the scale of what needed to be overcome. Lake's debacle lasted from 2 January to 10 April 1805 and is covered in detail in the author's own work on the Second Maratha War. The author of *Observations of the Attack on Mud Forts* summarised the siege:

> We have seen a detachment of two-thousand-five hundred sepoys sent to besiege a Fort, 1,700 yards in circumference, defended by a garrison of

---

35   George Wrottesley (ed.), *The Military Opinions of General Sir John Fox Burgoyne* (London: R. Bentley 1859), p.306.
36   James Young, *Galloping Guns* (London: Leonaur Books, 2008), p.232.

upwards 5,000 men. With this detachment, a train of artillery was sent consisting of two brass and two iron 18 pounders. The brass guns r[an] and became unserviceable; and with the two iron 18 pounders, a breach had to be effected in two different ramparts. Such was the scarcity of military stores, that ammunition could not be spared to demolish the defences, lest enough should not remain to effect a breach; and so scanty was the supply of implements in the engineer's department, that they were not procurable for the working parties. There were not even bamboos sufficient to make a second set of scaling ladders. The first set was lost in an unsuccessful assault; and before another could be made, it was necessary to send for bamboos upwards of a hundred miles.[37]

Indeed, after the first few assaults, things had looked so uncertain that a brief negotiation had ensued, but Lake had been adamant that the Jats surrender totally, and his attitude was echoed by Captain-Lieutenant Young:

If Runjeet Singh … effects to ride the high horse … and indeed does not humble himself to our superiority as siege Lords, to whom he owes fealty and submission, we should prefer enduring any hardship of climate, any loss of war to sinking in the minutest degree, in that opinion of the natives, by which alone, we may be said to rule India.[38]

Nothing better exemplifies the spirit and outlook of the British towards their place in India than this one statement from a junior officer of artillery. Indeed, the *Firangis* with their cannons felt themselves the lords of India because there was no place they could not take, that is, until Bharatpur. The damage this did to the reputation of the British was potentially significant, and had it occurred before the principal Maratha Confederates had been defeated in the Deccan and Hindustan in 1803, or in the immediate aftermath of Holkar's siege of Delhi in 1804, the damage might have been irreversible.

However, it must be remembered that Runjeet Singh was alone. Holkar had returned with a few allies and had added to the miseries of the Grand Army by raiding their supply lines on the road to Agra, but the Jat Rajah's money and supplies were not inexhaustible, and his people were suffering tremendously. The Marathas, being the last great power capable of resisting the company, were too politically fractured, militarily battered and too far away to make anything from Lake's folly, and as such, the British were able to receive a massive payment from Bharatpur and then withdrew to run Holkar to ground once again.

Pester outlined the state of the army, which had undertaken the operation in April, just before the siege ended:

Situated as we were, all our battering guns rendered useless nearly, by continual and almost incessant firing, our heavy shot completely

---

37 Anon., *Observations on The Attack of Mud Forts*, p.6
38 Young, *Galloping Guns*, p.177.

expended, nearly one- third of our officers and men killed and wounded (the Infantry, for the Cavalry were never within range of the shot), and those who remained, worn out almost by constant severe duty; under all these calamities, a peace, an honourable one, was to us an object desirable to be obtained. Our troops had now, for five complete months, been exposed continually to all the hardships and fatigues attending sieges in this fatal climate; constantly distressed by the scorching beams of a vertical sun in the trenches by day, and watching with that vigilance necessary to be observed before an enemy by night.[39]

For many, it was not just that the British had been repulsed but that they had been unable to achieve their goal despite assaults, and it was this that made the siege truly singular. Pester, too could sense the implications for the Company's reputation when he wrote, 'British arms in this part of the world had never before experienced such a check and a loss so severe, and without carrying our point.'[40]

As Burgoyne noted, to attack a place in an accelerated fashion, the general in question required great powers of judgement and the utmost security in his supplies and means. The author of the *Observations on the Attack of Mud Forts* agreed that, though certainly capable of taking any fort or fortress, a new attitude was required:

Of seven storms [on Mud Forts] he [the author] had seen five unsuccessful, [which tallies with the first assault on Sasni and the four on Bharatpur] in which upwards of 120 British Officers and 3,000 were killed or wounded, [casualties of Bharatpur] he had himself in the same service, but survived, not to partake of the glory, but to lament, with others, the humbled reputation of the army ... if our mode of warfare, on such service, be not improved ... we shall have nothing to expect but defeat, disaster and disgrace. To experience these must wound the pride of every British soldier, but to be defeated by those upon whom he has been accustomed to look with contempt, is truly mortifying.[41]

Based on Sasni and Bharatpur, he determined that there had never been enough care given to the approach to reducing mud forts, especially if they were protected by strong ditches and it is worth the reader's time to explore his thoughts more fully regarding how best learn from the Bengal Army's bruising experiences in Hindustan.

Conclusions

The siege of Seringapatam brought the British to the pinnacle of their reputation as 'siege lords' in South Asia. The conquest of Mysore was something typically Mughal

---

39   Pester, *War & Sport*, p.329.
40   Pester, *War and Sport*, p.386.
41   Anon., *Observations on The Attack of Mud Forts*, pp.III–IV.

in the relentless efficiency of bringing the siege train to the site of the fortress, progressing the attack and completing the breach, whereas in the next five years no matter the size of either the target or the garrison the British tended to throw themselves at it in the inevitable expectation of finding a way in by dint of the courage and audacity of their battalion officers rather than science.

In their quest to appear as a sort of supermen who were irresistible no matter the opposition, the army of the EIC, and especially that of the presidency of Bengal, were able to achieve great feats of arms, but strayed into lax practices that could have completely altered their reputation in India. The encouragement to cut corners in the face of what was considered an inferior enemy was, of course, driven by circumstances. The Jat campaign, for instance, was an offshoot of an unforeseen emergency, and after he had gained some momentum, Lake, when confronted with the very gates of an enemy by whom he felt wronged, could not fail to attempt it in much the same manner as Pester had done in the ceded provinces.

Despite this, the lessons of the period between 1792 and 1805 were not so much well learned and adapted as allowed to become dormant. The fall of the Maratha state removed the necessity to progress the Company style beyond what it had become, and the necessity to reduce and storm fortified places declined sharply, as in their absence, there was no state powerful enough to oppose the British on equal terms until the Sikh Wars.

When fortified positions that appeared too archaic for prudence once more became targets, the King's and Company officers continued to default to aggression rather than method as is exemplified in 1814, when Major General Rollo Gillespie attempted to storm the Nepalese fortifications of Nalapani during the Gurkha War, an act which saw his first three attacks repulsed leaving Gillespie himself dead, after which the British bombarded the fort for five days, forcing the Gurkhas to evacuate the ruins.[42]

Some lessons were learned, especially after a pretext was found to settle the score with the Jats. Bharatpur was, of course, not invincible to a commander who respected its strength, for in 1825, the British returned under Lord Combermere, and this time they knew what not to do. Instead, they conducted an operation similar to that of the last siege of Seringapatam. Here, the British wisely secured the sluice gates of the reservoir that had been used to flood the great ditch in 1805, allowing the besiegers to properly invest the place, and to bombard and lay mines, which widened the breaches to the extent that the fortress was carried in two hours.

With the fall of Bharatpur in January of 1826, the British could once again reassert their reputation, so colourfully framed by Captain-Lieutenant Young. The reputation of the Siege Lords would remain untested and unchallenged for another 30 years, until 1857, when the mutiny of the Bengal Army incited a war that once again revealed how precarious the EIC's position in India really was, and how vital the British ability to storm all before them remained.

---

42 See, John Pemble, *Britain's Gurkha War: The Invasion of Nepal, 1814–1816* (Barnsley: Frontline Books, 2008)

Bibliography

Allen, W.H., *An Illustrated Handbook of Indian Arms* (London: William H. Allen & Co, 1880)

Anon., *Observations on The Attack of Mud Forts* (Calcutta: Pereira, 1813)

Clive, Henrietta, Nancy K Shields (ed.) *Birds of Passage: Henrietta Clive's Travels in South India 1798-1801* (London: Eland, 2016)

Creighton, J.N., *Narrative of the Capture of Bhurtpore in the province of Agra, upper Hindustan, by the forces under the command of Lord Combermore, in the latter end of 1825, and beginning of 1826* (London: Privately Printed, 1830)

Duncan-Jones, Caroline M. (ed.), *Trusty and Well Beloved: The Letters Home of William Harness, an Officer of George the Third* (London: S.P.C.K, 1957)

Fraser, James Ballie, *Military Memoir of Lt-Col James Skinner, CB* (London: Smith Elder & Co, 1851)

Martin, Montgomery, *The Despatches, Minutes and Correspondence of the Marquess of Wellesley, K.G. During His Administration in India* (London: John Murray, 1836)

Muir, Rory, *Wellington: The Path to Victory* (New Haven: Yale University Press, 2013)

Pemble, John, *Britain's Gurkha War: The Invasion of Nepal, 1814-1816* (Barnsley: Frontline Books, 2008)

Provan, Joshua, *Bullocks Grain and Good Madeira: The Maratha and Jat Campaigns, 1803-1806* (Warwick: Helion and Company, 2021)

Sardesai, Govind Sakharam, *New History of the Marathas* (Bombay: Pheonix Publications, 1946)

Wrotteesely, George (ed.), *The Military Opinions of General Sir John Fox Burgoyne* (London: R. Bentley 1859)

Young, James, *Galloping Guns: The Experiences of an Officer of the Bengal Horse Artillery During the Second Maratha War 1804-1805* (London: Leonaur Books, 2008)

6

The Bombardment of Antwerp, February 1814

Andrew Bamford

In the range of options available by which to prosecute a siege, bombardment comes to mind primarily as an act of desperation or malice. One thinks of the British at Copenhagen in 1807, where time pressures prevented a more conventional siege and storm,[1] or, in an earlier age, the Russian burning of Cüstrin in 1758 or Frederick the Great's 1760 bombardment of Dresden. However, bombardment could also be used to target a purely military objective, as was demonstrated by the British-led operations against Antwerp in February 1814. Here, the goal was to destroy the French fleet sheltered in the city's dock-basin. Antwerp itself could not be taken by siege with the forces available, and so bombardment was seen as a next-best alternative by which the warships could be destroyed. This chapter offers a narrative history of this relatively short operation, providing a day-by-day account that draws on official reports as well as personal correspondence and memoirs of those who were there. As the author has observed elsewhere, this brief campaign furnished material for a disproportionate number of eyewitness accounts by British participants:[2] this allows this relatively minor operation to be examined in considerable detail from the perspective of the commanders who planned it, the engineers and artillerymen who executed it, and the infantry who shivered in the covering siegeworks. As well as providing useful technical and personal details, these testimonies reveal differing views regarding the level to which collateral damage and civilian casualties were deemed to be acceptable in a siege scenario – although, that said, conclusions on this head need in part to be drawn from the fact that many of the participants managed to avoid consideration of that issue in their accounts.

This is not the place to explore the background to this campaign in detail, but a certain amount of context is required to understand the pressures and limitations that the British commanders were operating under in their unfamiliar role

---

1 See Thomas Munch-Petersen, *Defying Napoleon: How Britain Bombarded Copenhagen and Seized the Danish Fleet in 1807* (Stroud: Sutton, 2007), in particular pp.193–199; John Harding Edgar, *Next to Wellington: General Sir George Murray* (Warwick: Helion, 2018), pp.78–89.
2 Andrew Bamford, *Triumphs and Disasters: Eyewitness Accounts of the Netherlands Campaigns 1813–1814* (Barnsley: Frontline, 2016), p.ix.

as junior partners in a coalition force. Upon the collapse of Napoleon's forces in Germany at the end of the 1813 campaign, something of a power vacuum existed in that portion of the Low Countries that had once formed the United Provinces or Dutch Republic, and which, after successive satellite incarnations as the Batavian Republic and Kingdom of Holland, had been directly annexed to France in 1810. Partisans of the exiled House of Orange were not slow to take advantage of the situation and were supported by the arrival of allied flying columns. For Britain, these developments presented both an opportunity and a challenge: on the one hand it was a key British war aim that the Dutch coast should be in friendly, or, at worst, neutral, hands; on the other, any claim for a stake in the future of the area required redcoated troops on the ground if it was to be taken seriously and the British Army was already stretched perilously thin. In due course, however, a weak force of four brigades plus supporting cavalry and artillery was scraped together from garrison and depot forces and sent out under the orders of General Sir Thomas Graham. Graham was newly returned from the Peninsula, where his last duties had been to oversee the bloody siege of San Sebastian.[3]

Initial British land operations were confused, as contact was made with the Dutch, Russian, and Prussian forces operating in the area, and efforts were made to agree a common plan of campaign. With conflicting goals and lacking any direction from their notional commander-in-chief, Crown Prince Karl Johan of Sweden, the erstwhile *Maréchal* Jean Bernadotte, relations between the allies were not easy. Eventually, Graham was able to obtain the agreement of *Generalleutnant* Friedrich Wilhelm, Graf Bülow von Dennewitz, commanding the Prussian III Armeekorps, to cooperate in operations against Antwerp. In doing so, he was working towards one of the major tasks that he had been given, as outlined on 4 December 1813 by the Secretary of State for War and the Colonies, Lord Bathurst. Having stressed that 'the main object in sending to Holland the British Force under your command is to give your Countenance and Support to the Exertions made by the People of Holland to vindicate their National Independence', Bathurst now drew Graham's attention to:

> … another Object in which the British Interests are deeply involved. I mean the destruction of the Naval Armaments at Antwerp. If at any time you should find it possible by marching suddenly on Antwerp to occupy such a position as would enable you to destroy the ships which it is understood are now laid up there, you should perform an essential service to your Country.

Bathurst acknowledged that 'A lengthened operation for that purpose is neither compatible with the description of the force under your Command nor with the Service under which you are to be employed under my former instructions', but

---

3   For a more detailed background, see Andrew Bamford, *A Bold and Ambitious Enterprise: The British Army in the Low Countries, 1813–1814* (Barnsley: Frontline, 2013), pp.6–57. On war aims, see also Rory Muir, *Britain and the Defeat of Napoleon 1807–1815* (New Haven: Yale, 1996), pp.241–261, 280–298, and, with a focus on Anglo-Dutch relations, G.J. Reynier, *Britain and the Establishment of the Kingdom of the Netherlands 1813–1815: A Study in British Foreign Policy* (London: George Allen and Unwin, 1930), pp.57–116.

emphasised that if allied cooperation could be secured, then such an operation might become feasible. Just to make sure that there was no misunderstanding, Bathurst reiterated 'that it is the destruction of the Naval Armament, not the capture of the Citadel or Town which should be the principal object of your exertions.'[4]

Unfortunately for Bathurst's hopes, the initial allied advance of 10–13 January 1814 did not produce any meaningful results. Snowy and icy weather hampered the approach march, and some stiff fighting took place as the Prussians drove back the French covering forces to the southeast of Antwerp. When the British came into action on 13 January, they achieved a small battlefield success in clearing the village of Merxem, directly outside Antwerp itself, but were then obliged to withdraw in order to conform to Prussian movements. Bülow having subsequently established that the fears of a French counterattack against his left flank were groundless – the French field forces having instead withdrawn – a second combined advance was agreed upon and on 2 February the allied forces were once more before the walls of Antwerp where Graham's redcoats again cleared Merxem of its defenders and thereby secured the positions from which the subsequent siege operations would take place.[5]

The remainder of this chapter, being focussed on the bombardment and not the wider operation, will not deal further with the activities of Bülow and his corps, but it should nevertheless be emphasised that without the Prussian covering force the British siege operations would not have been able to proceed. Indeed, as we shall see, it was the eventual departure of the Prussian troops that helped precipitate the end of the British efforts, although, as we shall also see, that was by no means the decisive factor that Fortescue's *History of the British Army* painted it to be. That, however, is to get ahead of the story, and we must now return to the situation as it stood in the late morning on 2 February with the Second Battle of Merxem petering out as the remaining French forces withdrew into Antwerp.

Objectives and Means

As Bathurst's instructions made clear, Graham's objective was not Antwerp itself, but the fleet moored in the city's harbour basin. Beginning with the handful of semi-derelict survivors of the old Dutch fleet, this had been built up by 1814 to number 22 sail of the line, with a further nine on the stocks. As early as 1809, the naval build-up had been deemed sufficient a threat to make Antwerp the target of the 'Grand Expedition' that ultimately bogged down in the fever-swamps of Walcheren; since then, the French had been kept confined to the Scheldt by a blockading squadron of the Royal Navy. Although hindsight tells us that the fact that the ships were of limited utility due to constructional defects, this was not known at the time. Furthermore, while the difficulties of using Antwerp as an operational base, due to the difficult navigation of the lower Scheldt, were known, this was a double-edged sword – while

---

4 TNA: WO1/199, pp.41–43, Bathurst to Graham, 4 December 1813.
5 Summary from Bamford, *A Bold and Ambitious* Enterprise, pp.79–141.

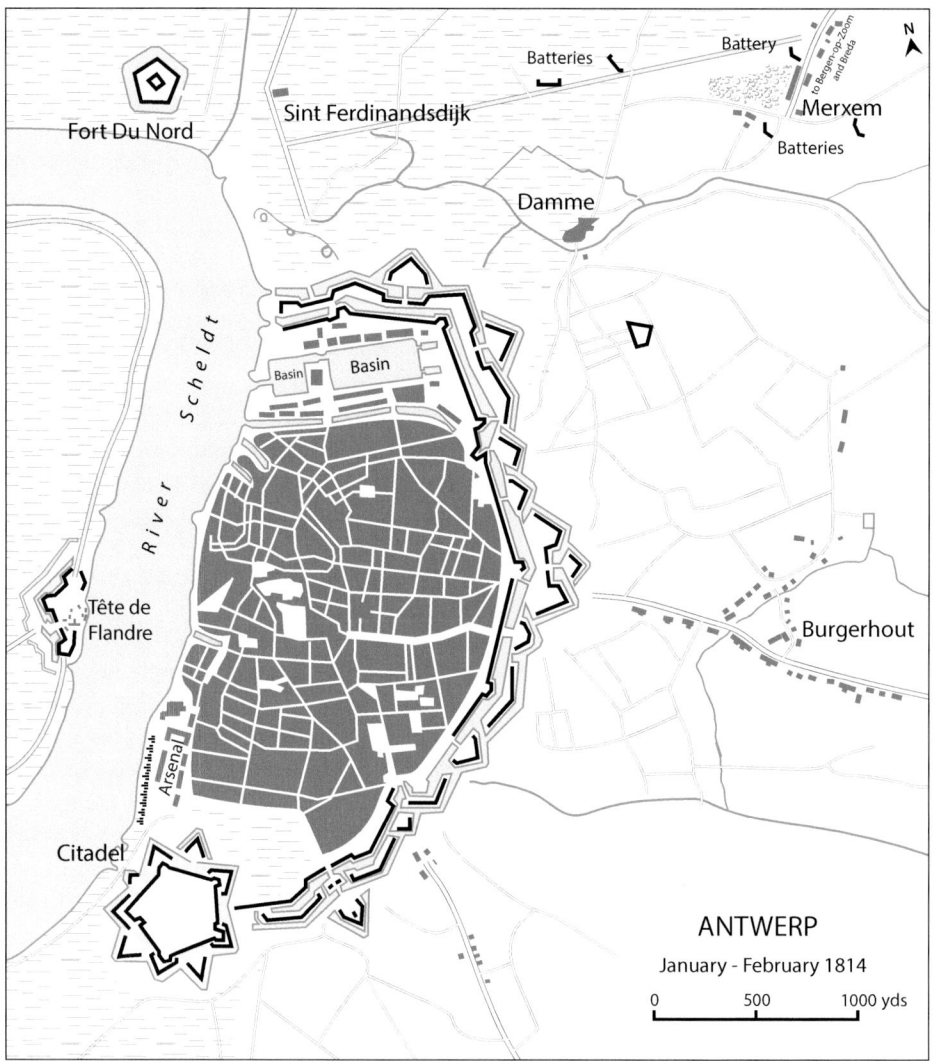

ANTWERP
January - February 1814

it would be difficult for the French to get out, it was also difficult for the British to make any sort of naval assault up the river.⁶

Such, indeed, was the contemporary importance placed on the capture or destruction of the ships that it was even possible to use this incentive to help secure

---

6 W.M. James, *The Naval History of Great Britain During the French Revolutionary and Napoleonic Wars* (London: Richard Bentley, 1837), vol.VI, p.259; Rif Winfield and Stephen S. Roberts, *French Warships of the Age of Sail 1786–1861. Design, Construction, Careers and Fates* (Barnsley: Seaforth, 2025), pp.32, 49, 59, 95–96. For a discussion of the Napoleonic building programme in the wider context of Britain's concerns over Antwerp, see Paul Leyland, 'Antwerp: Britain's Achilles Heel', in Nicholas James Kaizer (ed.), *Sailors, Ships and Sea Fights: Proceedings of the 2022 'From Reason to Revolution 1721–1815' Naval Warfare in the Age of Sail Conference* (Warwick: Helion, 2024), pp.175–198.

the Prussian cooperation that made the operation possible. Major General Herbert Taylor, serving as a brigade commander under Graham while on detachment from his regular duties as Secretary to Queen Charlotte, but here acting as Graham's envoy, was able to hold out to Bülow in the planning stages of the operation the promise that, 'in the event of the Capture of the French ships, the Prize Money to be distributed to the troops concerned British & Prussian would be very considerable, & that even if we should only succeed in burning them a very handsome gratuity would be made to the Troops'. As Taylor went on to report to Graham, the Prussian general 'appeared to receive this information with Pleasure'.[7]

The basin in which the ships were moored was situated to the north of the city but fell within its main line of defensive works. The abortive operations of 13 January had at least allowed the British to get a good look at Antwerp and its fortifications and to plan accordingly. The commanding officer of Graham's Royal Engineer detachment, Lieutenant Colonel James Carmichael Smyth, having seen his target first hand, reported that:

> I have no hesitation in giving it as my decided opinion that whenever circumstances will allow us to advance in front of Antwerp for 48 hours we can burn the Naval Arsenal and the fifteen sail of the line now moor'd within the Docks or great Basins lately excavated. The Ships lay with their broadsides touching each other, or nearly so. There are several most advantageous Positions for Mortars & Howitzers within so short a range as 1500 and even 1300 yards.[8]

To the north of the main fortifications, Fort du Nord on the banks of the Scheldt provided additional cover, although with only 'six 12pdrs and two small Howitzers' its ordnance was unable to materially impact against British operations centred around Merxem.[9] The forts lower down the estuary that defended against an assault from the sea had been bypassed by the allied overland advance. Insofar as the positions outlined by Carmichael Smyth as necessary for a bombardment were concerned, to establish them required the seizure of the ground around the villages of Merxem and Damme (the latter lying even closer to Antwerp than the former), and along the Sint Ferdinandsdijk which ran from Merxem out towards Fort du Nord over ground that was otherwise inundated as an additional defensive measure. The French had developed plans for additional outworks in this sector but had been unable to do more than mount a few fortress guns in the villages and along the Sint Ferdinandsdijk, which were overrun during the infantry fighting. The required area was therefore in British hands at the close of the fighting on the 2nd, although it could be brought under fire by the heavy guns emplaced on Antwerp's ramparts; these had inflicted casualties on Graham's attacking infantry during the fighting around Merxem, and would continue with counterbattery and harassing fire throughout the coming operation.

---

7   TNA: WO1/199, pp.565–568, Taylor to Graham, 31 January 1814.
8   TNA: PRO30/35/6, pp.7–11, Carmichael Smyth to Mann, 14 January 1814.
9   TNA: PRO30/35/1, pp.64–67, 'Memorandum respecting the proposed Operations against Antwerp' by Carmichael Smuth, 7 January 1814.

Now that the bulk of the effective field forces in the area had been driven away from Antwerp during the January operations, allied intelligence appreciations placed little store by the capabilities of the remaining garrison forces holding the city. Prussian intelligence, passed on to Taylor during his conference with Bülow and relayed by Taylor to Graham, reported that, '[A] deserter from Antwerp stated his information to be that in general that Provisions were very scarce, & the inhabitants loud in their complaints of being deprived of their stock for the Garrison. That the Deficiency of Gunners was such that the Batteries were manned by Conscripts.'[10]

Other intelligence of a more practical kind was obtained by Carmichael Smyth, thanks to the desertion to the allies of 'A Lieut. Colonel and a Captain of Engineers the latter of who has been stationed at Antwerp for two Years (Dutchmen by birth) [who] came over to us during the Affair of Yesterday [13 January] & they have promised to add to a plan of Antwerp (which I already have) all the French Improvements'.[11] Both men confirmed Carmichael Smyth's belief that burning the fleet was an entirely realistic prospect, and the Dutch-born captain, Jan van Gorkum, would become a useful confidant and friend to the British engineer. Their report also allowed the allies an accurate appreciation of the strength of the garrison:

18 Battns, 450 each: 8100
Two Companies of Sappers: 180
Artillery: 600
With Sailors and Veterans the whole were completed to amount to 10,000 Men. 7/8 of the Soldiers might be reckoned Conscripts.[12]

However, just as the January operations allowed the British to improve their knowledge of their target, so too did they provide the French with a warning of likely allied intentions. Carmichael Smyth certainly recognised this, noting in his memorandum on the proposed February operations that: 'The five Bastions and three Ravelins counting from the Scheldt and by the Red Gate were only mounted with 20 Guns – 6pdrs, 8pdrs and 12pdrs when Capt. Gorcum left Antwerp. There were no Mortars or Howitzers. As however there were plenty in the Town it is to be supposed that these Points are now completely armed.'[13]

In this assumption he was entirely correct, with the corrective measures due in great part to the arrival at the end of January of Antwerp's new governor, Lazare Carnot. Brought out of retirement and hastily commissioned *général de division*, the old Revolutionary 'Organiser of Victory' revitalised the defence both in terms of morale and practical measures.

---

10   TNA: WO1/199, pp.565–568, Taylor to Graham, 31 January 1814.
11   TNA: PRO30/35/6, pp.7–11, Carmichael Smyth to Mann, 14 January 1814.
12   TNA: PRO30/35/1, pp.64–67, 'Memorandum respecting the proposed Operations against Antwerp' by Carmichael Smuth, 7 January 1814.
13   TNA: PRO30/35/1, pp.64–67, 'Memorandum respecting the proposed Operations against Antwerp' by Carmichael Smuth, 7 January 1814.

Miles Byrne, an officer of the 3e Régiment Étranger (formerly the Légion Irlandais and still with a number of Irishmen among its commissioned ranks), gave an account of Carnot's leadership as perceived by the defenders:

> Carnot's presence alone was equal to a reinforcement of troops; it both encouraged the soldiery and imposed on the immense population, which could not with safety be entrusted at this critical moment. He soon proved by his genius and firmness that the town could resist for more than six months, and he accordingly desired that the inhabitants should make provisions for that length of time, or leave the town, whichever they liked best. A great many got the provisions necessary, others chose to leave the town and crossed the river. Several of the latter had sad reason to repent of the step they had taken, as they were plundered by the Cossacks encamped on the other side of the river, and they found no protection from those pretended liberators.[14]

With respect to the artillery defences, Carnot recognised that the British attacks had been against the 'weakest part of the fortifications'. Accordingly:

> [H]e soon had a battery of 36-pounders and several mortars erected on the rampart. On that part, by the aid of thousands of the inhabitants whom he put in requisition to make small sacks and fill them with clay and carry them to the rampart, the parapets of the battery were all constructed during the night. On the 1st [sic] of February, 1814, the English, no doubt, must have been surprised to find from this weak point, as they thought, a battery of 12 pieces of 36-pounders, and four great mortars, playing on their works, which prevented them advancing.[15]

Whereas Carnot had the resources of an arsenal at his disposal, Graham and his artillery commanders had rather less on which to draw when it came to assembling a siege train. Five companies of artillerymen had been sent out with Graham's army, numbering 795 rank and file on 25 January 1814.[16] Two companies had been equipped as field brigades leaving the remainder available for siege work. If men were available, however, sufficient heavy ordnance was not. Back in December, Carmichael Smyth had been able to inform Lieutenant General Gother Mann, Inspector-General of Fortifications, to whom he sent detailed reports throughout the campaign, that although he was still awaiting engineer stores: 'We have however in the mean time six Transports with a small Battering Train, originally meant for Spain and from which we will derive great assistance consisting of 6 24 pdrs, 6 Howitzers, 4 Mortars, & 4 68pdr carronades, with the requisite Platforms and Stores.'[17] Platforms, to enable the pieces to be emplaced for siege work, were provided for all these guns,

---

14   Miles Byrne, *Memoirs of Miles Byrne* (Dublin: Maunsell and Co., 1907), vol.II, p.146
15   Byrne, *Memoirs*, vol.II, p.147.
16   TNA: WO17/1773, Monthly Return of 25 January 1814.
17   TNA: PRO30/35/6, pp.2–6, Carmichael Smyth to Mann, 31 December 1813.

and sufficient for a further eight 24-pounders in addition, although the hammers and augurs required to assemble the platforms were missing.[18]

This train, such as it was, was brought forward during the January advance, but moved a day behind the infantry and so was not available to go into action on 13 January, even had the circumstances permitted it.[19] For the February advance, the carronades – of no use for a long-range bombardment – were left behind but 'To make up for the want of our own artillery, all the serviceable Dutch mortars, with all the ammunition that could be collected, were prepared at Willemstadt'. This need to employ locally-obtained weapons was due to the fact that, as Graham reported, 'the state of the weather for some time back not only prevented my receiving the supplies of ordnance and ordnance stores from England, but rendered it impossible to land much of what was on board the transports at Willemstadt, the ice cutting off all communication with them'.[20]

The total available ordnance stood as follows:

English Ordnance { 4 10" Mortars
2 8" Howitzers
6 24 Pounders

Dutch Ordnance { 3 12" Gomers Mortars
4 11" Mortars
6 7½" Mortars[21]

This gave a total of 12 British and 13 Dutch pieces, or 25 in total. A Gomers or Gomer mortar is one possessed of a conical chambre in the breach, named for its inventor, the eighteenth-century French artillery officer Louis-Gabriel, Comte de Gomer. The three listed here were very likely modern pieces of French manufacture, as the conical chamber had been adopted in 1791 as a modification to the 12-inch mortars of the existing Gribeauval system.[22] The diary of Major James Stanhope, Graham's aide-de-camp, seemingly confirms this by referring to them not as Gomers but as Napoleons, but also adds the vital detail that there was very little suitable ammunition for them.[23] The remaining pieces, meanwhile, were genuinely Dutch in origin, old, and of dubious quality.

---

18 TNA: PRO30/35/6, pp.7–11, Carmichael Smyth to Mann, 14 January 1814; this also specifically identifies the howitzers as being of 8-inch calibre.
19 TNA: PRO30/35/6, pp.7–11, Carmichael Smyth to Mann, 14 January 1814.
20 TNA: WO1/199, pp.569–578, Graham to Bathurst, 6 February 1814.
21 TNA: WO1/199, pp.569–578, Graham to Bathurst, 6 February 1814.
22 Marc Morillon, 'The Siege Mortars and Their Related Skills during the Napoleonic Era', *Napoleon Series*, <https://www.napoleon-series.org/military-info/organization/c_mortars.html>, accessed 13 February 2022.
23 Gareth Glover (ed.), *Eyewitness to the Peninsular War and Waterloo. The Letters and Journals of Lieutenant Colonel the Honourable James Stanhope 1803 to 1825* (Barnsley: Pen and Sword, 2010), p.136.

On the face of it, even leaving aside issues of ammunition supply and ordnance quality, this seems like a woefully inadequate siege train for such a crucial mission and begs the question of why a veteran professional engineer like Carmichael Smyth was so sanguine about the prospect of success. Part of the answer may lie in an optimistic outlook that also led him to champion the ultimately disastrous attempt to take Bergen-op-Zoom by escalade, and which at Antwerp clearly caused him to grossly underestimate the effectiveness of French countermeasures. However, he was also working from a very limited background of recent professional experience that would have equally affected any of his professional colleagues. The simple matter was that the British Army and its associated Ordnance services had very little experience, recent or historic, in directing artillery of any form against warships. The last time that British soldiers had destroyed significant amounts of enemy shipping without direct naval assistance would seem to have been at St Servan in 1758, but that had been achieved by an overland assault force that had seized the harbour from the landward size and then torched the contents – clearly not an option that was remotely practicable here.[24] Subsequent bombardments of shipping – Rodney at Le Havre in 1759, Nelson's failed 1801 attempt on the invasion flotilla at Boulogne – had been naval affairs, and even with Ordnance personnel aboard the bomb vessels there was simply no 'lessons learned' system to filter back the fruits of that experience, such as it was.[25]

Two final points should also be made to head off hindsight criticism of the optimism shown by Graham and Carmichael Smyth as to what could be achieved with such apparently limited resources. Firstly, one might consider that when the option was considered to destroy the Danish fleet by bombardment during the 1801 Copenhagen expedition – an option which was ultimately not acted upon due to the agreement of an armistice after Nelson's attack on the blockships covering the approaches, but which in conception comes closest to what was being attempted at Antwerp – the Royal Navy assigned seven bomb vessels mounting 14 mortars, a fraction of what was employed in 1807 when the whole city was targeted.[26] This demonstration of what would seem to be an equal if not greater level of optimism would tend to suggest that it was not Carmichael Smyth who was uniquely ignorant of what could be achieved with limited resources, but Britain's scientific military community more widely. Secondly, and perhaps more importantly, one ought also consider that, even if Carmichael Smyth or Graham did have misgivings about the means at their disposal, no additional resources were available. As with Wellington's failed siege operations at Badajoz in 1811 and Burgos in 1812, the options were either to make the attempt with what was available or not make the attempt at all.

24 Anon., *British Minor Expeditions 1746–1814* (London: HMSO, 1884), p.11.
25 For the employment of Ordnance personal to man the mortars on bomb vessels, see Francis Duncan, *History of the Royal Regiment of Artillery* (London: John Murray, 1873), pp.83, 85.
26 Ole Feldbæk, *The Battle of Copenhagen 1801* (Barnsley: Pen and Sword Maritime, 2016), pp.130, 208–211; ship details from 'Battle of Copenhagen, 2nd April 1801' at <https://threedecks.org/index.php?display_type=show_battle&id=152>, accessed 15 March 2024; on 1807 ordnance, see Munch-Petersen, *Defying Napoleon*, pp.200–201.

2 February

Sir Thomas Graham's operational orders for the attack on Merxem made it clear that as soon as the village had been secured, it was to be fortified and held as the base for siege operations. Specifically:

> As soon as the Enemy is driven out of the Village, the avenues leading into it from Antwerp will be strongly occupied and Lieut. Colonel Carmichael Smyth Commanding the Engineers will lose no time in forming Barricades and such other Works as he may think necessary to defend it against any attempt by the Enemy.
>
> As it may be expected that the Enemy will throw shells from the Ramparts of Antwerp, the Village will at first be occupied with all attention to find shelter for the troops from the effect of such Bombardment, no more being allowed to remain in the Village and within range of the Artillery of the place than what are necessary for its defence and for carrying on the works already mentioned.[27]

Merxem was occupied by the Brigade of Guards under Colonel Lord Proby – composed of the 2nd battalions of all three Foot Guards regiments – which enabled the troops who had stormed the village to withdraw and regroup.[28] As the Guards marched in, it became clear that the village had been abandoned by the French but not by its inhabitants, who drew down the sympathy of Corporal Alexander Frederic Meuller of the 1st Foot Guards:

> Upon our arrival in Marksam [sic], the doors and window-shutters were closed, and we, of course, thought the place had been deserted; but no sooner did the wary inhabitants know who were masters, than they came out with liquors to treat the victors. Some of our soldiers looked upon the poor creatures with contempt, calling them deceitful; but the more reflecting rather pitied them, having the misfortune to be so near the seat of war.[29]

Also with the Guards was Ensign Thomas Slingsby Duncombe of the Coldstreamers, who 'was on a covering party immediately after we arrived' while the rest of the brigade went to work on the batteries. 'They shelled us a good deal', he wrote in his diary, before relating how 'I was then detached with fifty men to a post on the

---

27 TNA: PRO30/35/6, pp.15–16, Copy of the Disposition for the Attack on Merxem by General Sir Thomas Graham.
28 Duke of Wellington's Regiment Museum (DWRM): Diary of William Thain, entry for 2 February 1814. For a complete order of battle for Graham's troops during the Antwerp operation, see appendix to this chapter.
29 Alexander Frederic Meuller, 'Selections from Letters, Written from the Netherlands, By A.F. Meuller, Corporal in the First Foot Guards', reproduced as an appendix to Anon., *Journal of a Soldier of the Seventy-First Regiment Highland Light Infantry, From 1806 to 1815* (Edinburgh: Balfour and Clarke, 1822), pp.192–228, quoting p.200.

right of the village, to defend our right, and there I found a snug house; but the balls rattled about it a good deal'.[30]

The engineer stores that were necessary to put the village of Merxem into a defensible state, and subsequently to construct the battery positions necessary for the bombardment, were the responsibility of Lieutenant John Sperling of the Royal Engineers, who had charge of around 100 locally-sourced waggons – 'most unfit for their present object' – and a detachment of the Royal Sappers and Miners. Sperling's letters home to his father, which he reclaimed after the latter's death and subsequently published, show him to have been a serious-minded officer with strong religious convictions. He was evidently well-thought-of by Carmichael Smyth, who gave him a series of responsible and important jobs, even when more senior Royal Engineer officers were available, and who later employed him as a personal aide during a post-war tour of local fortifications. The stores under Sperling's charge comprised 'a quantity of timber adapted for the laying of platforms for artillery, making magazines, and whatever may be required for the construction of field batteries, a quantity of sand-bags, entrenching tools and so forth'.[31] The engineer train moved in the rear of the Prussian troops, and was initially accompanied by the siege mortars before these were sent on ahead with the rest of the artillery, under the charge of Lieutenant Colonel Gold, Royal Artillery. The morning of 2 February found Sperling and his stores at Wuustwezel, around 12 miles short of Merxem, and they waited there until noon when orders were received to move forward: 'Being prepared, and only waiting instructions, the waggons were put in motion, and we arrived at Merxem about seven in the evening. The stores were deposited around the Church'.[32]

As a large and central structure, possessed in its bell tower of an ideal observation post, Merxem's church became the central point of the British occupation of the village and was used throughout the operation by Graham and his staff. Taking an initial view, the anonymous eyewitness responsible for an epistolary history of the campaign offered a favourable opinion of what could be seen: 'From the spire of the village church we had a most extensive view of the city and suburbs, and of every ship and vessel in the harbour, as well as in the river; and it appeared next to an impossibility that those, particularly in the outer bason, could escape being destroyed'.[33]

With Merxem secure, Graham's instructions required that 'Lieut Colonel Smyth will likewise mark out such Mortar Batteries as he may think necessary and will

---

30 Thomas H. Duncombe (ed.), *The Life and Correspondence of Thomas Slingsby Duncombe, late MP for Finsbury* (London: Hurst and Blackett, 1868), vol.I, p.14, diary entry of 2 February 1814.
31 John Sperling, *Letters of an Officer of the Corps of Royal Engineers, From the British Army in Holland, Belgium, and France, to his Father, From the Latter End of 1813 to 1816* (London: John Nisbet, 1872), p.22.
32 Sperling, *Letters*, p.25.
33 Anon., *Letters from Germany and Holland, during the years 1813–14; with a detailed account of the operations of the British army in those countries, and of the attacks on Antwerp and Bergen-op-Zoom, by the Troops under the command of Gen. Sir T. Graham* (London: Thomas & George Underwood, 1820), p.132. The author of this account appears to have been a member of Graham's staff, but it has not been possible to establish their identity.

send in to the Adjutant General a detail of such working parties as he may require'.[34] Naturally, the engineer had already made plans to site the batteries that would be required for the bombardment. Working through the options prior to allied advance, he had initially considered a more methodical approach by which the outlying villages of Eeckeren and Wilmarsdonk and the ruins of Fort Pimentel at Oorderen, all to the north of Merxem in an area now mostly replaced by modern docks, would be used to establish batteries from which the bombardment could begin while also providing a base for securing more advanced positions along the Sint Ferdinandsdijk. This, however, would have placed an area of inundations between the besieging troops and the remainder of the allied forces and so made communication difficult. An similarly advantageous site for the batteries, Carmichael Smyth concluded, 'may be found equally in front of Merxem without the disadvantages of detaching by avoiding which our Corps can be well Kept together to repulse any Sorties or other operations the Enemy may undertake', and he went on to assert that: 'The occupation of Merxem seems the Key to our operations, and in front of it, and on its left, there appear to be several situations where Mortar Batteries may be placed to great effect so as to take the Basin diagonally as also in its greatest length.'[35]

So much for the theory; putting this into practice meant hard work under enemy fire, as related by Lieutenant Sperling who, having delivered his stores to the village, was assigned with his detachment of sappers to the construction of one of the mortar batteries, heading one of four such working parties. The position assigned to Sperling was out along the Sint Ferdinandsdijk; this meant that it was one of the most exposed locations, but that the workers, who also included a 'working party of the line' under the direction of Sub-Lieutenant Thomas Adamson of the Royal Sappers and Miners, benefited from being able to repurpose the existing French works. As Sperling explained: 'Ferdinand's Dyke extends from the river nearly to Merxem, affording complete cover from the artillery on the side away from the town. To remedy this they were constructing the work, which in part now served us for a parapet, and brought us much nearer to the docks than any of the other batteries in progress.'[36]

Even with the benefit of the existing French works as a basis, however, the task was still a substantial one and the work would continue throughout the night, as it did across the front of Graham's lines. At least in Sperling's sector, however, no attempt was made by the defenders to impede the night-time operations, which were covered by detachments from the light infantry units in Graham's army – the understrength 2/52nd Light Infantry and a composite rifle battalion containing companies from all three battalions of the 95th. With the former unit was Lieutenant Charles Shaw, who described the unenviable conditions along the Sint Ferdinandsdijk:

---

34   TNA: PRO30/35/6, pp.15–16, Copy of the Disposition for the Attack on Merxem by General Sir Thomas Graham.
35   TNA: PRO30/35/1, pp.64–67, 'Memorandum respecting the proposed Operations against Antwerp' by Carmichael Smuth, 7 January 1814.
36   Sperling, *Letters*, p.26.

> The weather being frosty, and the ground of a damp and spongy nature, the only way to get shelter from the shot was by digging holes in the dyke. Tools being put into our hands this was speedily done; but, as the wet continued dropping from the upper part of the holes, and froze as it fell upon us, our position was anything but pleasant. I was in the same hole with my brother subaltern, now Major M'Dowall of the 44th, and I well remember we lay on each other in turn in order to keep our bodies warm; meanwhile, the enemy kept up a very sharp fire of shot and shell.[37]

Private Thomas Morris of the 2/73rd, who was on an outlying picquet, confirmed that 'it was a beautiful night, but very cold'.[38]

3 February

In Graham's report of the operations against Antwerp, he informed Bathurst that: 'No time was lost in marking out the batteries which by the very great exertions of the artillery under Lieut.-Colonel Sir George Wood, and the Engineers under Lieut.-Colonel Carmichael Smyth, and the good-will of the working parties, were completed and armed by half-past three of the 3rd.'[39]

'Exertions' and 'good-will' were all very well, but the actual work for those on the ground was altogether more taxing, not to mention dangerous, than Graham's brief report suggests. Sperling, out along the Sint Ferdinandsdijk, reported that: 'With the dawn the interior slope of the parapet had reached its height; this was built up of sand-bags, but the upper part had not yet attained sufficient thickness to resist artillery'. The daylight exposed Sperling's party to fire from the guns of Fort du Nord, which rendered proceedings altogether riskier and more extended:

> The work was continued as much under cover as possible, but Sapper Northam losing his leg, and some other casualties having taken place, the men were withdrawn altogether from the exterior. The labour was now carried on under the shelter of the parapet, the men bringing the mould from the rear in baskets; thus not advancing so rapidly but in comparative security, the shot passing over us or lodging in the parapet. In the meantime the carpenters were laying down the platforms for the six mortars, and an excavation was made for a powder magazine.
>
> By noon the battery was completed, after sixteen hours' labour. Resigning it into the hands of the Artillery, I brought off the working party.[40]

---

37 Charles Shaw, *Personal Memoirs and Correspondence of Colonel Charles Shaw KTS &c* (London: Henry Colburn, 1837), vol.I, p.25. The other officer referred to was Ensign D. McDowell, 2/52nd Light Infantry.
38 Thomas Morris, *Recollections of Military Service, in 1813, 1814, & 1815, through Germany, Holland, and France, including some Details of the Battles of Quatre Bras and Waterloo* (London: James Madden, 1845), p.96.
39 TNA: WO1/199, pp.569–578, Graham to Bathurst, 6 February 1814.
40 Sperling, *Letters*, pp.26–27.

Whereas Sperling was directing operations, closer to Merxem Private Morris was getting his hands dirty:

> Our regiment was engaged, in the face of the enemy, in constructing a sandbag mortar battery, which is formed as follows: – Such a number of men, as may be deemed sufficient, are provided each with a canvas bag, which is to be filled with sand and secured at the mouth by a string. These are then deposited in rows, under the superintendence of the artillery-men, and in an hour or two a battery may be so formed, which will bear a great deal of battering.[41]

Sand, if it could be found, might have been easy enough to dig, but the earth itself was 'frozen as hard as stone' according to newly-commissioned Ensign Richard Master of 2/1st Foot Guards, who had charge of one of his battalion's working parties.[42]

While all this was going on, the necessity remained for the work to be covered by the infantry outposts. Duncombe, for one, was engaged, being sent 'to attack an enemy's picquet in a wood, but we found there was treble our number, so we returned without firing a shot at them, though they kept up a brisk fire against us'.[43]

Graham's brief consideration of the construction of the batteries may in part have been due to the fact that his attentions that morning were distracted by matters of strategy. Notwithstanding earlier assurances of Prussian cooperation, Graham in the early hours of the morning received a communication from Bülow to the effect that the Prussian corps was ordered to push forward towards Malines and Brussels, where it already had outposts, and from there to cooperate in the main allied advance on Paris. As Herbert Taylor recorded it, the Prussian commander 'expressed his anxiety to know when our batteries would begin to play upon the fleet, that he might make his arrangements for his own movement after affording us every aid and co-operation in his power'. Accordingly, Taylor was sent to liaise with the Prussians and convey Graham's request that their movement be postponed until 6 February in order to allow the bombardment to take place. To this, Taylor was soon able to obtain Bülow's agreement that he would remain 'until our business was done and everything brought off down to the last platform'.[44]

Returning from his liaison duties, Taylor was able to join Graham in the tower of Merxem's church, which had been appropriated as a vantage point from which the bombardment could be observed. They were joined there, at least briefly, by HRH Prince William, Duke of Clarence, who was on the continent in order to dodge his creditors and find a bride, and who had accompanied the army in a private capacity. This is not the place to explore the various myths that place the future William

---

41 Morris, *Recollections*, pp.97–98,
42 Sir David Fraser (ed.), 'An Ensign at War. The Narrative of Richard Master, First Guards', *Journal of the Society for Army Historical Research*, vol.LXVI, no.267 (Autumn 1988), pp.127–145, quoting p.132.
43 Duncombe (ed.), *Thomas Slingsby Duncombe*, vol.I, p.14, diary entry of 3 February 1814.
44 Ernest Taylor (ed.) *The Taylor Papers. Being a Record of Certain Reminiscences, Letters, and Journals in the Life of Lieut.-Gen. Sir Herbert Taylor GCB, GCH* (London: Longmans Green, 1913), p.138, Taylor to the Duke of York, 4 February 1814.

IV in the thick of the fighting around Antwerp – some of which even erroneously have him present at the January battle instead – but suffice it to say that this was the closest that the royal personage actually came to enemy fire.[45] Morris, indeed, remarked that 'a shot from the enemy struck the steeple, and gave him an intimation that his royal person was not exactly safe', after which the Duke took himself to the rear.[46] Taylor confirms that the steeple was hit multiple times, and Duncombe recorded that the Guards 'were fairly shelled out of the village' by the fire from Antwerp, so it is hardly to be wondered that the church was hit.[47]

The fire, when it opened between 3:30 and 4:00 p.m., was brisk – 'very hot on both sides',[48] said Duncombe, who had found a windmill from which to watch the show – but it did not do much immediate execution. Taylor felt that the British shot was falling short, at least at first, but towards 5:00 p.m. the correct range had been found and the firing began to have an effect, with fires being observed around the dockyard; specifically, according to Taylor, in the 'Magasins de Comestibles, and a church this side of the arsenal'.[49] Thus heartened, the gunners kept up their work until dusk, and, indeed, continued a little after it. Ensign William Thain was with the 33rd, which, along with the rest of Taylor's Second Brigade, was deployed 'exactly in rear of the batteries, ready to support the men in the trenches should the enemy have made a sortie'. From there, he had 'a fine view of the firing, which was beautiful after it became dark. No rockets were made use of but the shells were filled with the same combustible matter that they used'.[50]

Clearly, the initial results were not those that had been hoped for, with limited damage done to the dockyard area and no apparent harm to any of the warships. As Graham reported, even the firing in the short window of decent light on the evening of the 3rd made it clear that 'the defective state of the Willemstadt [sic] mortars and ammunition was too visible'.[51] The only solution was to move the defective mortars to a position further forward, but this would eat into the time available, now strictly limited by the agreement with Bulow to have the operation completed by the 6th. On the other hand, the delay would enable the 24-pounders to be properly brought into play, as the battery position intended for them – sited further forward, and under fire from French outworks – was still being finished even as the mortars began their bombardment. It is not clear whether the 24-pounders actually fired on the 3rd; they are listed by Graham as amongst the ordnance available on that day, but the account of Sperling, who was assisting with the construction of the battery, suggests that the work went on until dusk and makes no mention of the guns having been placed in battery.[52] Carmichael Smyth's account, on the other hand, states that the

---

45 For an essay critiquing the Clarence legends, see Bamford, *Triumphs and Disasters*, Appendix IV, pp.208–224.
46 Morris, *Recollections*, p.97.
47 Taylor (ed), *Taylor Papers*, p.139, Taylor to the Duke of York, 4 February 1814; Duncombe (ed.), *Thomas Slingsby Duncombe*, vol.I, p.14, diary entry of 3 February 1814.
48 Duncombe (ed.), *Thomas Slingsby Duncombe*, vol.I, p.14, diary entry of 3 February 1814.
49 Taylor (ed), *Taylor Papers*, p.139, Taylor to the Duke of York, 4 February 1814.
50 DWRM: Diary of William Thain, entry for 3 February 1814.
51 TNA: WO1/199, pp.569–578, Graham to Bathurst, 6 February 1814.
52 Sperling, *Letters*, pp.27–28.

24-pounders did fire on the 3rd, along with 17 mortars and two howitzers (which matches Graham's figures), but his account also states that the fire was opened in the morning, which rather suggests some confusion with the events of the 4th and implies that the long guns were not, therefore, employed until that day.[53]

4 February

The night of 3–4 February was again cold, Morris recollecting that the men of his battalion 'began to dig caves to keep off a portion of the cold air, as well as a shelter from the shot. The ground being chiefly sand, we were enabled to do this with some pickaxes and shovels, the loan of which we obtained from the sappers and miners'.[54] Firing by the French resumed at around 8:00 a.m.,[55] but so far as the British were concerned the priority for the 4th was to re-work the arrangement of the batteries, including the re-siting of the Dutch mortars, and efforts to this end continued throughout the day although from noon this was accompanied by a resumed bombardment by the available ordnance.

Carmichael Smyth's report is silent on the details of the engineering work, but the activities of Lieutenant Sperling, who seems to have had an incredibly busy day, give some indication of the volume of tasks that was necessary. His initial job was:

> [T]o take a working party to make some splinter-proofs for the picquets or advanced sentinels, and to give some additional protection to the 24-pounder battery. The splinter-proofs were formed by throwing up earth against the walls of outbuildings on the side toward the town, and laying sloping beams of timber against the other side; thus a place of tolerable security was obtained.[56]

This work was carried out with the corpses of those killed the previous evening still strewn around the battery position. During the afternoon, Sperling was employed in the rear, directing repair work for platforms that had been damaged by the firing of the ordnance mounted on them. This job took him to dusk, and only then was he shifted to the work – which might have been assumed to have been a more significant priority – of establishing new mortar batteries in suitably advanced positions for the outranged Dutch pieces, further along Sint Ferdinandsdijk from the initial positions. Moving the batteries forward also rendered them more vulnerable, and required additional defensive precautions: 'The place for the mortars was cut in the Dyke, the mould excavated being thrown on the side towards Fort du Nord, to which we were now nearer than before'.[57] Sperling was relieved at 1:00 a.m. on the 5th, and, as he put it 'ventured to undress' in the hope of getting some much-needed

---

53   TNA: PRO30/35/6, pp.12–14, Carmichael Smyth to Mann, 5 February 1814.
54   Morris, *Recollections*, pp.98–99.
55   Timing from DWRM: Diary of William Thain, entry for 4 February 1814.
56   Sperling, *Letters*, p.29.
57   Sperling, *Letters*, p.29.

sleep. Three hours later, however, he was called back to the work, which was not finally completed until midday on the 5th.

With suitable emplacements not available for the old Dutch pieces, only two of these – both of the smaller 7½-inch type – fired during the course of the 4th, the main burden of the high-angle fire being kept up by the four British mortars and the three 12-inch Gomers. These were supported by the two British howitzers and by the six 24-pounders, with four of the latter being employed to fire heated shot. However, notwithstanding the efforts to further reinforce their battery position, by the end of the day only a single 24-pounder was still in action. At least two had burst, always a potential risk with heated shot, while others had been disabled by the heavy French counterbattery fire. Two would be repaired in time to fire the following day, but the other three were rendered unusable for the remainder of the operation.[58]

The results of the day's firing appear to have been negligible, Taylor giving a summary in his letter home to the commander-in-chief:

> Our shells and red-hot shot have been thrown with great precision, but until 3, when I left Merxem, without much effect, although we clearly saw them fall into the shipping and arsenal; but I suspect all have been sodded and otherwise made bomb-proof; and we feel the want of rockets most seriously. The town was on fire to the left of that building (the arsenal), but there being little wind it probably would not spread.[59]

Thain in his diary was more succinct: 'No flames or even smoke from the fleet'.[60]

For most of the infantry, now with little to do unless working parties were required, thoughts turned to comfort. Battalions that had spent the previous day and night in forward positions were rotated to the rear; those posted in Merxem and in support of the batteries tried to make the best of things as they prepared for another clear and freezing night. Duncombe summarised the day as follows: 'Went on picquet again under a very heavy fire; lost a few men; laid down under the bank; the firing ceased towards night; procured a cold chicken and some brandy about nine, and got through the night pretty well, but was excessively cold'.[61]

5 February

With time running down towards the deadline imposed by the impending departure of the Prussians, 5 February had become the decisive day on which the bombardment must either succeed or fail. The three 12-inch Gomers remained available, as

---

58 TNA: WO1/199, pp.569–578, Graham to Bathurst, 6 February 1814; for the detail re. heated shot, see TNA: PRO30/35/6, pp.12–14, Carmichael Smyth to Mann, 5 February 1814. With respect to the burst guns, see Taylor (ed.), *Taylor Papers*, pp.139–140, Taylor to the Duke of York, 4 February 1814.
59 Taylor (ed.), *Taylor Papers*, p.139, Taylor to the Duke of York, 4 February 1814.
60 DWRM: Diary of William Thain, entry for 4 February 1814.
61 Duncombe (ed.), *Thomas Slingsby Duncombe*, vol.I, p.14, diary entry of 4 February 1814.

did the British ordnance, less the three disabled 24-pounders. Of the older Dutch pieces, the big 11-inch mortars remained unemployed, but all six of the 7½-inchers were brought back into action in their new positions, albeit that no beds were available for them. Graham reported that: 'The fire was kept up all day. The practice was admirable, but there was not a sufficient number of shells falling to prevent the enemy extinguishing fire whenever it broke out among the ships, and our fire ceased entirely at sunset'.[62] In short, the bombardment had been a failure, and time was now up.

It was clear to onlookers that things were not going as hoped. Duncombe's summary of the day's events was terse: 'part of our village on fire, and some of our guns dismounted; things going on very badly; marched the men up to their cantonment; the batteries were opened again, but they said to no use, as they never hurt the fleet'.[63] Claiming to feel unwell, Duncombe visited his battalion's surgeon, who diagnosed an ague and sent him to the rear. This cannot have made him popular with his fellow subalterns, as Master observed that the Guards battalions were already short of junior officers, with those present obliged to do double duty without relief.[64]

Major General Taylor provided rather more detail in his correspondence with the Duke of York, concentrating, as befits an infantry brigadier, on the measures taken to protect his men from the increasingly heavy French fire:

> Our fire opened again early yesterday and was kept up from twelve mortars and howitzers (which had been partly moved during the night more to the right) and three 24-pounders, with great vivacity and precision, every shot or shell almost falling into the basin or arsenal, but without visible effect on the ships, although they must have suffered very much.
>
> The enemy's fire was also very heavy early in the day, but relaxed some time after, clearly from their attention being taken up with the fires in the shipping, which they were extinguishing, numbers being discerned supposed to be so employed.
>
> Towards the afternoon their fire on Merxem became so heavy, some houses being on fire, and shells came in so thick that after having sixteen men of the 78th wounded by the explosion of one shell in a house, I ordered my brigade, which was on duty, to move out to the rear, out of reach of this destructive fire, and kept them out till dark, when I relieved the 35th pickets with the 33rd, and moved the rest behind the dyke, where I continued with them all night until relieved at daylight by part of the first division.[65]

Ensign Thain was in one of Taylor's battalions, and also experienced the 'excessively hot' fire. Notwithstanding Taylor's decision to pull his men back, however, elements of his brigade were soon back in the firing line as Thain went on to note that 'After dark our Brigade went into the trenches, when our Grenadiers were employed in

---

62   TNA: WO1/199, pp.569–578, Graham to Bathurst, 6 February 1814.
63   Duncombe (ed.), *Thomas Slingsby Duncombe*, vol.I, p.15, diary entry of 5 February 1814.
64   Fraser (ed.), 'An Ensign at War', pp.132–133.
65   Taylor (ed.), *Taylor Papers*, p.140, Taylor to the Duke of York, 6 February 1814.

dragging the guns and mortars out'.[66] Sperling was also involved in superintending the process of dismantling the siege works, telling his father that 'At eight o'clock in the evening took charge of a party to collect the tools and light materials from the different batteries, to have them packed in the wagons ready for moving. This was accomplished soon after midnight'.[67]

Although conscientious officers like Thain and Sperling continued to do their duty, it seems likely that the failure of the bombardment caused a dip in morale that led others to neglect theirs. We have already seen the feckless Duncombe absent himself on a slim pretext; now Captain Edward Fitzpatrick, serving with Thain in the 33rd, appeared 'in a state of intoxication on the public parade, or place of assembly of the 33d Regiment … when the regiment was preparing to go into the trenches, under St. Ferdinand's Dyke, and in front of the enemy'.[68] There was seemingly an attempt to resolve this matter within the regiment, but after Fitzpatrick turned up drunk for duty a second time, there was no avoiding a court martial, held on 25 April, at which he was sentenced to be cashiered. Quite possibly, though it is hardly adequate defence, he had been drinking to keep warm; if so, he was not the only one. We have already seen Ensign Duncombe resort to brandy against the cold, and from the ranks Private Morris remembered that 'Our daily supply of schnapps we found of considerable service; it was served out in camp kettles, each man receiving about a third of a pint, which they generally put out of sight at once, in order to make sure of it'.[69] It is perhaps not surprising that Sperling noted the need for punishment for drunkenness amongst the Sappers and Miners in the aftermath of the Antwerp operations,[70] and it may have been in a vain attempt to insure against such problems, as well as his stated aim of providing 'something warm to take early by way of breakfast in the cold damp climate', that Graham had earlier requested of Horse Guards 'the sending out of a large quantity of Cocoa ready pounded'.[71]

Withdrawal and Conclusions

With the successful extraction of the siege artillery from the works, active operations came to an end, for, as Graham reported, 'the Enemy did not show himself out of their works, so that every thing was quietly withdrawn w'out any molestations'.[72] At least in some quarters, this was aided by deception: Sergeant Thomas Jackson of 2/Coldstream Guards recalled that 'fires were left burning, and some effigies of soldiers set up in the batteries'.[73] Considering the heavy French counterbattery fire,

---

66   DWRM: Diary of William Thain, entry for 5 February 1814.
67   Sperling, *Letters*, p.31.
68   Charles James (ed.), *A Collection of the Charges, Opinions, and Sentences of General Courts Martial* (London: T. Egerton, 1820), p.634; for more on this see also Bamford, *Triumphs and Disasters*, pp.200–202.
69   Morris, *Recollections*, p.96.
70   Sperling, *Letters*, pp.32–33.
71   TNA: WO1/199, pp.339–342, Graham to Bathurst, 31 December 1813.
72   TNA: WO1/199, pp.593–595, Graham to Bathurst, 7 February 1814.
73   Eamonn O'Keeffe (ed.), *Narrative of the Eventful Life of Thomas Jackson: Militiaman and Coldstream Sergeant, 1803–1815* (Solihull: Helion, 2018), p.60.

losses of personnel during the siege operations – as distinct from the initial storming of Merxem – had been fairly light: three rank and file killed, three officers, three sergeants, two drummers, and 48 rank and file wounded. The bulk of the casualties were incurred by the Royal Artillery, heavily exposed throughout the operations, and the 2/78th due to the unlucky shell described by Taylor. Equally unlucky were two of the three officer casualties, Lieutenant Robert Stowers and Ensign George Chapman of the 2/37th, who each lost a leg to the same cannonball;[74] the third officer casualty, only slightly wounded in a separate incident, was Ensign Alexander Redock of the 2/44th. Sixteen artillery horses were killed, nine wounded, and a further 12 recorded as missing; regarding the latter, it was recorded that they had escaped during the firing, Graham's Deputy Adjutant General, Lieutenant Colonel Arthur Macdonald, felt obliged to add the qualification that 'They were Dutch Horses'.[75]

On the other hand, it was not immediately clear what, if anything, had been achieved by the bombardment. However, two days after the British withdrawal one Adrian P. Hayes, of Rotterdam, deserted from the French artillery at Antwerp and was subsequently interrogated by Graham's Commandant of Headquarters, Captain Charles Hamilton Smith. Hayes had been wounded in the hand serving the guns near the 'Windmills of Dam' – by inference, this must have been on 2 February – and reported that there were more than 400 in Antwerp's hospital, to the extent that extra paillasses had to be provided. As Hamilton Smith summarised it, Hayes reported that: 'The shells did the greatest mischief the first day; and one ship was on fire – the two following days they fell almost all in a street to the left of the basin. He believes that six or seven shells only fell amongst the ships – 5 inhabitants & 2 galley slaves were killed – no soldiers lost their lives.'[76] There was no obvious damage to any of the ships, but the garrison 'are all very much discouraged – provisions bad & scanty'.

This report confirmed what Graham had already emphasised in his report to Bathurst, that 'there was not a sufficient number of shells falling to prevent the enemy extinguishing fire whenever it broke out among the ships'.[77] Carmichael Smyth, writing home to his superior, gave more details but came to essentially the same conclusion:

> It was impossible to sustain the most sanguine hopes of a favourable Result, notwithstanding however the utmost Exertions of every Officer and Man in the Army and the very excellent Practice made by the artillery who threw more than 2000 shells & a continued fire of three Days we have not obtained the objective in view. The ships were repeatedly on Fire, several Fires Kindled in Buildings round them, and a large Store Room containing

---

74 As related in, amongst others, Glover (ed.), *Stanhope*, p.135, although Sergeant Jackson reported that the injuries were caused by the explosion of a shell – see O'Keeffe (ed.), *Eventful Life of Thomas Jackson*, p.59.
75 TNA: WO1/199, p.589, Casualty Returns 3–5 February 1814.
76 TNA: WO1/199, pp.601–602, 'Information received from Adrian P. Hayes of Rotterdam who deserted from the artillery at Antwerp 8th February'.
77 TNA: WO1/199, pp.593–595, Graham to Bathurst, 7 February 1814.

the Biscuit and Provisions for the Fleet being destroyed by being put on Fire. The Vessels themselves, however, have not been bunt and I am afraid it is not possible to destroy them without much larger means than we at present possess. The Enemy kept a number of People on board of these provided with Buckets and Fire Engines, and as fast as a fire was kindled it was extinguished. It appears therefore absolutely necessary to ensure success that we should not only have Mortars enough to set fire to the shipping but to enable us to throw such an overwhelming quantity of Shells as to render it totally impossible for any men to work whilst exposed to it. Congreave Rockets have been requested as peculiarly adapted for this sort of Service but as I have never seen them used I cannot give an opinion. We have none with this Army, or they would unquestionably have been made us of, and had a fair trial.[78]

Carmichael Smyth was not the only observer to wish that rockets had been available for the bombardment – Taylor, as we have seen, remarked upon their absence and the anonymous writer of *Letters from Germany and Holland* says that as late as 6 February 'we were anxiously looking out for the arrival of the rockets'[79] – but as a professional engineer officer his desire to see them employed is particularly interesting. However, it is hard to reconcile their use against a precision target – in relative terms at least, as opposed to the city as a whole – with their notorious record for inaccuracy. Carmichael Smyth might not have seen these weapons in use, but must have been aware of the consequences of their employment at Copenhagen in 1807, where they contributed both to the extensive fires – though their role in this relative to the mortars may be overstated – and to the psychological cowing of the defenders.[80] That said, there is no guarantee that the results would have been quite what was hoped for even had rockets been available end employed: in the near-contemporary siege of Danzig, in which neither the French defenders nor their Russian attackers showed much concern for collateral damage, Russian observers were decidedly underwhelmed by the effects of a rocket bombardment.[81]

Insofar as moral judgements about the use of such weapons are concerned, it is perhaps telling that even after Copenhagen the government was criticised for mounting the attack at all, not for the manner in which the attack was carried out.[82] Nor does there appear to have been criticism of the fires started ashore by rockets during the 1806 test of rockets against shipping at Boulogne.[83] Judging by the silence of the majority of commentators on the option of a more intensive bombardment that might have led to civilian casualties, the employment of rockets against Antwerp would have raised few eyebrows in Graham's army. Precedents from the

---

78 TNA: PRO30/35/6, pp.12–14, Carmichael Smyth to Mann, 5 February 1814.
79 Anon., *Letters from Germany and Holland*, p.134.
80 Munch-Petersen, *Defying Napoleon*, pp.200–202.
81 See account of Ensign Rafail Zotov, in Darrin Boland (ed.), *Recollections from the Ranks: Three Russian Soldiers' Autobiographies from the Napoleonic Wars* (Solihull: Helion, 2017), p.117
82 Munch-Petersen, *Defying Napoleon*, pp.233–235 242–246.
83 A.D. McCaig, '"The Soul of Artillery": Congreve's Rockets and their Effectiveness in Warfare', *Journal of the Society for Army Historical Research*, vol.78, no.316 (Winter 2000), pp.252–263.

Peninsula indicate that Wellington certainly considered collateral damage through the employment of rockets acceptable if the situation was urgent – as at Santarem in November 1810, and as might be argued in the case of Graham at Antwerp due to time pressure – but unacceptable otherwise.[84]

Yet burning Antwerp to the ground was certainly no part of Graham's plan, and would have been disastrously counterproductive in terms of Britain's long-term aims in the region. Interestingly, this seems to have been more readily understood in the ranks than by more senior commentators, with Sergeant Jackson recording the understanding that the fleet was targeted in lieu of the city due to a request from the Prince of Orange that the civilian population be spared.[85] Graham may also have been mindful of the need to avoid the shocking destruction – including extensive fires – that had accompanied his capture of San Sebastian, and an analysis of his correspondence during the Flanders operations suggests that events in Spain were still very much in his mind and may have influenced his decision-making.[86]

When, in March, Graham did have rockets available to use, these were tried against an individual French warship that had dropped down from Antwerp to bombard British positions on the lower Scheldt. It proved impossible to hit the target, reinforcing the dangers of using these weapons for precision work, but the threat was enough to compel the French captain to disengage.[87] There was never any thought of using this weapon against Antwerp or any other fortress, although as late as 25 March, even after the Bergen-op-Zoom debacle had cost him much of his infantry, Graham was still hoping that a renewed operation against Antwerp might be possible if only the Royal Navy, the Dutch, the Saxons, and other allies could be persuaded to cooperate, and if only more engineers and sappers could be found – Carmichael Smyth produced a memorandum that called for an additional 22 Royal Engineer officers and 210 Sappers and Miners, an increase of 350 percent on the forces then at his disposal![88] There was, in reality, about as much chance of this operation ever being mounted as there was of Carnot, that honest old Revolutionary, accepting the million-Franc bribe to hand over the fortress that it was proposed to offer him via a British agent in the aftermath of the failed bombardment.[89]

Ultimately, the bombardment of Antwerp failed for much the same reasons that limited the success of Graham's operations more generally: that is to say, he had not been provided with sufficient means to achieve his objectives in the time available. British attempts to make up the shortfall by dragooning unwilling allies in to make up the numbers – and then blaming them when they preferred to follow their own

---

84   David Saunders, '1617 "The Soul of Artillery"', Notes and Documents, *Journal of the Society for Army Historical Research*, vol.79, no.320 (Winter 2001), pp.342–346.
85   O'Keeffe (ed.), *Eventful Life of Thomas Jackson*, p.58.
86   See correspondence in TNA: WO1/199–201: one might speculate that the preference for a bombardment at Antwerp and an escalade at Bergen-op-Zoom both stemmed from a desire to avoid a repetition of the slaughter in the breech at San Sebastian.
87   This action is covered in more detail in Bamford, *Bold and Ambitious Enterprise*, pp.241–244.
88   TNA: WO1/200, pp.367–378, Graham to Bunbury, 25 March 1814; p.385 Memorandum of forces needed for attack on Antwerp, 25 March 1814.
89   TNA: WO1/199, pp.671–672, J.M. Johnson to Graham, 18 February 1814, pp.673–675. Graham to Johnson, 19 February 1814.

national interests – have coloured much of our understanding of this campaign as a whole and the Antwerp operations in particular. Until the present author began to research and write about these campaigns in the 2010s, the most substantial English-language treatment of the British Army's service in the Low Countries during 1813 and 1814 was that of Fortescue in his *History of the British Army*, and with the relevant volume having been prepared in the aftermath of the Great War the Prussians are unsurprisingly cast as scapegoats.[90] Fortescue implies that Bülow left Graham in the lurch when in fact Graham fully realised that the bombardment was a futile exercise, as is clear from his letters in its aftermath, but more junior figures on the British side did seize upon Bülow's departure as a cause for failure and this popular contemporary perception may have provided Fortescue with his cue.[91]

Put simply, however, the movements of the Prussians were immaterial, and the British lacked the necessary material resources to carry out a bombardment of the intensity necessary to destroy the French fleet. Yet, had those resources been available, then it is very likely that the result would have been to flatten or incinerate large portions of a city that was intended by British politicians to become part of a favourably inclined client state in the post-war settlement, without any guarantee of destroying the well-protected warships. In this failure to think things through, as with the neglect through hurry of other essentials to the prosecution of the campaign, the British government gave the British Army an impossible task.

So, what lessons or insights can be drawn from the experience of this short and ultimately futile operation? The main lesson is the point just made, that such operations require appropriate means – in terms of both equipment and time – to have a realistic chance of success. Within that, it should in particular be emphasised that in order to destroy a target by means of setting it ablaze, not only should a sufficient volume of fire be delivered to set it on fire in the first place but that this must be continued over a sustained period in order to thwart any attempts at firefighting. In terms of wider insights into the practice of siege warfare in this period, the remarkably low casualties suffered as a result of the French counterbattery fire stand out – particularly when contrasted to descriptions of its extent and ferocity – and emphasise the advantages provided by the use of the stone buildings of Merxem and the conveniently-aligned Sint Ferdinandsdijk as a means to shelter Graham's troops from both the fire and the elements. The latter point also reminds us that this was one of the coldest winters of the war – the operations coincided with London's last ever Frost Fair on the frozen Thames – and that simply sustaining an army in the field outside Antwerp was an achievement in itself and one that stands to the credit of the ordinary soldiers involved and whose voices and experience this chapter has sought to place on an equal footing with the technical aspects of the operation.

---

90   Hon. J.W. Fortescue, *A History of the British Army* (London: Macmillan, 1899–1920), vol.X, pp.11–12; see also, p.9, regarding the abortive January operations and referenced in the table of contents as 'Tricky behaviour of Bülow'.
91   See, for example, Fraser (ed.), 'An Ensign at War', p.133.

Appendix – Organisation of British Forces at the Bombardment of Antwerp

First Division: Major General George Cooke
    Guards Brigade: Colonel Lord Proby: 2/1st Foot Guards (708); 2/Coldstream Guards (479); 2/3rd Foot Guards (499)
    First Brigade: Major General John Byne Skerrett: 2/44th (399); 55th (295); 2/69th (433); Flank Companies of 2/21st and 2/37th (c.264)[92]
    Divisional Artillery: Rogers' Brigade
Second Division: Major General Samuel Gibbs
    Light Brigade: Lieutenant Colonel William Harris: 2/25th (319); 2/52nd (185); 54th (439); 2/73rd (474); Rifle Battalion (255)
    Second Brigade: Major General Herbert Taylor: 33rd (502); 2/35th (432); 3/56th (255); 2/78th (262)
    Divisional Artillery: Fyers' Brigade
Cavalry: Lieutenant Colonel Baron Linsingen
    2nd KGL Hussars (451, with 517 horses)
Unassigned Artillery: Lieutenant Colonel Sir George Wood[93]
    Truscot's Brigade; Tyler's Brigade; Hawker's Brigade

Unit strengths are given for effective rank and file as per Monthly Return of 25 January 1814, in TNA: WO17/1773.

Bibliography

Archival Sources
Duke of Wellington's Regiment Museum (DWRM)
    Diary of William Thain.
The National Archives, Kew (TNA)
    PRO30/35/1 Sir James Carmichael Smyth: Papers. Miscellaneous. Letters, observations, memoranda and reports. Entry book.
    PRO30/35/6 Sir James Carmichael Smyth: Papers. Netherlands and France. Dispatches. To Lieutenant General [Gother] Mann [Inspector General of Fortifications]. Entry books.
    WO1/199 War Office and predecessors: Secretary-at-War, Secretary of State for War, and Commander-in-Chief, In-letters and Miscellaneous Papers: Europe and the Mediterranean. Dutch Expedition (1813-1814). Commander's dispatches.

---

92 Calculated based on numbers of men returned as by these two battalions (the centre companies of which were in garrison) as 'on command'; that is, on duty away from the parent unit. Actual figure may have been slightly lower if detached men from the centre companies were also listed under this heading.
93 Total effective artillery strength amounted to 767, 737 of these being listed as 'on command' which in this context should be assumed to mean serving with one of the five active companies as opposed to with headquarters: no company/brigade level breakdown is available.

Published Sources

Anon., *British Minor Expeditions 1746-1814* (London: HMSO, 1884)

Anon., *Letters from Germany and Holland, during the years 1813-14; with a detailed account of the operations of the British army in those countries, and of the attacks on Antwerp and Bergen-op-Zoom, by the Troops under the command of Gen. Sir T. Graham* (London: Thomas & George Underwood, 1820)

Bamford, Andrew, *A Bold and Ambitious Enterprise: The British Army in the Low Countries, 1813-1814* (Barnsley: Frontline, 2013)

Bamford, Andrew, *Triumphs and Disasters: Eyewitness Accounts of the Netherlands Campaigns 1813-1814* (Barnsley: Frontline, 2016)

Boland, Darrin (ed.), *Recollections from the Ranks: Three Russian Soldiers' Autobiographies from the Napoleonic Wars* (Solihull: Helion, 2017)

Byrne, Miles, *Memoirs of Miles Byrne* (Dublin: Maunsell and Co., 1907)

Duncan, Francis, *History of the Royal Regiment of Artillery* (London: John Murray, 1873)

Duncombe, Thomas H. (ed.), *The Life and Correspondence of Thomas Slingsby Duncombe, late MP for Finsbury* (London: Hurst and Blackett, 1868)

Feldbæk, Ole, *The Battle of Copenhagen 1801* (Barnsley: Pen and Sword Maritime, 2016)

Fortescue, Hon. J.W., *A History of the British Army* (London: Macmillan, 1899-1920)

Fraser, Sir David (ed.), 'An Ensign at War. The Narrative of Richard Master, First Guards', *Journal of the Society for Army Historical Research*, vol.LXVI, no.267 (Autumn 1988), pp.127–145

Glover, Gareth (ed.), *Eyewitness to the Peninsular War and Waterloo. The Letters and Journals of Lieutenant Colonel the Honourable James Stanhope 1803 to 1825* (Barnsley: Pen and Sword, 2010)

Harding Edgar, John, *Next to Wellington: General Sir George Murray* (Warwick: Helion, 2018)

James, Charles (ed.), *A Collection of the Charges, Opinions, and Sentences of General Courts Martial* (London: T. Egerton, 1820)

James, W.M., *The Naval History of Great Britain During the French Revolutionary and Napoleonic Wars* (London: Richard Bentley, 1837)

Leyland, Paul, 'Antwerp: Britain's Achilles Heel', in Nicholas James Kaizer (ed.), *Sailors, Ships and Sea Fights: Proceedings of the 2022 'From Reason to Revolution 1721-1815' Naval Warfare in the Age of Sail Conference* (Warwick: Helion, 2024)

McCaig, A.D., '"The Soul of Artillery": Congreve's Rockets and their Effectiveness in Warfare', *Journal of the Society for Army Historical Research*, Vol.78, No.316 (Winter 2000), pp.252–263

Morillon, Marc, 'The Siege Mortars and Their Related Skills during the Napoleonic Era', *Napoleon Series*, <https://www.napoleon-series.org/military-info/organization/c_mortars.html>

Meuller, Alexander Frederic, 'Selections from Letters, Written from the Netherlands, By A.F. Meuller, Corporal in the First Foot Guards', reproduced as an appendix to Anon., *Journal of a Soldier of the Seventy-First Regiment Highland Light Infantry, From 1806 to 1815* (Edinburgh: Balfour and Clarke, 1822), pp.192–228

Morris, Thomas, *Recollections of Military Service, in 1813, 1814, & 1815, through Germany, Holland, and France, including some Details of the Battles of Quatre Bras and Waterloo* (London: James Madden, 1845)

Muir, Rory, *Britain and the Defeat of Napoleon 1807*-1815 (New Haven: Yale, 1996)

Munch-Petersen, Thomas, *Defying Napoleon: How Britain Bombarded Copenhagen and Seized the Danish Fleet in 1807* (Stroud: Sutton, 2007)

O'Keeffe, Eamonn (ed.), *Narrative of the Eventful Life of Thomas Jackson: Militiaman and Coldstream Sergeant, 1803-1815* (Solihull: Helion, 2018)

Reynier, G.J., *Britain and the Establishment of the Kingdom of the Netherlands 1813-1815: A Study in British Foreign Policy* (London: George Allen and Unwin, 1930)

Saunders, David, '1617 "The Soul of Artillery"', Notes and Documents, *Journal of the Society for Army Historical Research*, vol.79, no.320 (Winter 2001), pp.342–346

Shaw, Charles, *Personal Memoirs and Correspondence of Colonel Charles Shaw KTS &c* (London: Henry Colburn, 1837)

Sperling, John, *Letters of an Officer of the Corps of Royal Engineers, From the British Army in Holland, Belgium, and France, to his Father, From the Latter End of 1813 to 1816* (London: John Nisbet, 1872), p.22

Taylor, Ernest (ed.) *The Taylor Papers. Being a Record of Certain Reminiscences, Letters, and Journals in the Life of Lieut.-Gen. Sir Herbert Taylor GCB, GCH* (London: Longmans Green, 1913)

Winfield, Rif, and Roberts, Stephen S., *French Warships of the Age of Sail 1786-1861. Design, Construction, Careers and Fates* (Barnsley: Seaforth, 2025)

7

Revisiting British Accounts: The 'Other' Deadly Siege on the Northern Coast of Spain (1813)

Silvia Gregorio Sainz

Introduction

In 1813, as Arthur Wellesley, future Duke of Wellington, was preparing his next offensive into Spain, the port of Santander on the northern coast acquired a key role for the Anglo-Portuguese troops' advance.[1] Given the inadequate road infrastructure in the Iberian Peninsula, Wellington turned to the Royal Navy for logistics support, urging Captain Sir George Collier, commander of the naval squadron off the Biscay Coast, in April to transfer the army's supplies to the Cantabrian port and to tighten the blockade along that shore.[2] Santander, being close enough to the theatre of operations and capable of mooring transports in all-weather conditions, turned into a British depot.

It was thus essential to blockade the Spanish northern coast, and mainly the eastward ports in French hands, such as Santoña and San Sebastián, to safeguard communications and the safe delivery of supplies. Between them, the seaside town of Castro Urdiales (hereafter Castro) was controlled by a Spanish garrison, but not for long. Besieged by the divisions of *Généraux de division* Maximilien S. Foy, Jacques T. Sarrut and Giuseppe F. Palombini in May 1813, its fortress was stormed and, after a bloody resistance involving British participation, taken. It meant the destruction of the place and the annihilation of civilians. These operations, while apparently posing a problem to Wellington's army supply plans, contributed to diverting the attention of French forces from allied preparations behind the Portuguese border.

---

1 The information presented in this chapter is part of a broader research on Anglo-Spanish relations throughout the nineteenth century conducted as a member of the University of Oviedo research group 'OLE-6'.
2 John Gurwood (ed.), *The Dispatches of Field Marshall the Duke of Wellington K.G. during his various Campaigns in India, Denmark, Portugal, Spain, the Low Countries, and France. From 1799 to 1818* (London: J. Murray, 1852), vol.10, p.318, Wellington to Collier, 22 April 1813.

The episode has been an object of some research in Spain, especially after its bicentennial celebration, but without fully incorporating the British perspective into its narrative. Even if in the early 1900s the Spanish historian José Gómez Arteche had already integrated in his multivolume book the effective participation of a British squadron in the operations in Castro, based on primary sources, later national and regional historiography tends to oversimplify the allies' role, summarising it in barely a sentence.[3] However, back in 2013, in the context of my PhD research dealing with Anglo-Spanish relations in Cantabria during the Peninsular War, I found more significant information not only about the Spanish and British resistance in the Cantabrian port or the successful transfer of the garrison by the Royal Navy, but also about its takeover later in July. The initial results were then published in a volume entitled *La Guerra de la Independencia en Castro Urdiales. 11 de Mayo 1813*.[4]

This chapter aims to expand on that previously published work revisiting both the correspondence between the Government in London and the officers in the Peninsula (especially, the captains of the British squadron off Castro), and printed 'contemporary' sources such as John Marshall, Robert Southey, and W.F.P. Napier, among others.[5] Special attention is also paid to allusions to the episode in the Spanish and the British press, focusing, though, on the newspapers published in London.[6] The critical revision of these sources, following the New Historicism approach,[7] contributes by reconstructing the French siege of Castro in May 1813 from the British perspective to determine the actual role the allies played in the operations there, which are usually reduced to minor evacuation tasks in the Spanish discourse. It also looks at the impact the episode had on the British reading public, as well as on Wellington's actions in the broader context of the Peninsular War.

---

3   José Gómez Arteche, *Guerra de la Independencia. Historia Militar de España de 1808 a 1814* (Madrid: Depósito de la Guerra, 1902), vol.13, drew on Captain Robert Bloye's report. Meanwhile, José Muñoz Maldonado, *Historia política y militar de la Guerra de la Independencia de España contra Napoleón Bonaparte desde 1808 a 1814* (Madrid: José Palacio Ramos, 1833), vol.3, p.367, only mentioned Collier's participation in evacuation tasks; and, José Simón Cabarga, *Santander en la Guerra de la Independencia* (Santander: J. Simón Cabarga, 1968), p.249, briefly referred to British aid against French assaults in March 1813.
4   Silvia Gregorio Sainz, 'Asedio y destrucción de Castro Urdiales según las fuentes británicas: la participación del aliado británico en la defensa de la villa cántabra', in Miguel A. Sánchez Gómez (ed.), *La Guerra de la Independencia en Castro Urdiales. 11 de Mayo 1813* (Torrelavega: Ayto. de Castro Urdiales, 2015), pp.89–116.
5   The correspondence examined comes mainly from The National Archives (hereafter TNA), and the Archivo Histórico Provincial de Cantabria (hereafter AHPCan). British 'contemporary' source books include John Marshall, *Royal Naval Biography…* (London: Longman, 1829); Robert Southey, *History of the Peninsular War* (London: John Murray, 1832); and, W.F.P. Napier, *History of the war in the Peninsula and in the South of France, from the year 1807 to the year 1814* (Brussels: Meline, 1839).
6   Online accessed to British and Spanish papers via the websites of *The British Newspaper Archive* (hereafter BNA), <www.britishnewspaperarchive.co.uk>, *The Gazette* – Official Public Record, <https:// www.thegazette.co.uk> and the *Biblioteca Nacional de España – Hemeroteca Digital*, <www.bne.es/es>.
7   Stephen Greenblatt and Catherine Gallagher, *Practicing New Historicism* (Chicago: University of Chicago Press, 2000).

From Early Contacts to a Crucial Mission in 1812

After the Spanish uprising against Napoleon in May 1808 and the resulting alliance with Great Britain in July, the northern coast acquired a growing importance for the efficient development of both contending armies' operations in the area, or even further inland. The central position of modern-day Cantabria on the Bay of Biscay, together with its relatively good sea and land communications, turned the region into a point of special strategic interest, which the Government in London was soon aware of. From the outset of the Peninsular War, British agents in the area reported the strengths and weaknesses of the three main port towns along the Cantabrian coast: Santander, Santoña and Castro. Despite the neglected state of Spanish coastal defences at the beginning of the conflict, several factors explain these ports' advantageous characteristics: their suitable natural features, the adequate land infrastructures between them and Madrid, and their proximity to the main theatre of operations in key moments of the war. Their control was thus essential to serve as logistics hubs to guarantee the British blockade of the coast, easing communications and the safe reception and delivery of supplies, equipment and troops, and also to launch military operations to get hold of key territories and/or to divert enemy troops.

However, not all the Cantabrian ports carried the same weight; while British attention over Santander and Santoña was constant, the port of Castro only came into play on given occasions. Yet, its relevance was clear: placed right in the centre of the port network between San Vicente de la Barquera and San Sebastián, it was described as safer than that of Santander due to its topography and good defensive qualities. Built on a rocky promontory, Castro was protected both by the Cantabrian Sea and, on the southwest, by a medieval wall (up to 20–22 feet high and seven feet wide) framed by two batteries. The main town, including San Francisco and Santa Clara convents, stood immediately behind it and led to an upper area, which served as an inner fortification, formed by Santa Ana Castle-Lighthouse and Santa María Church. Thus, like the French redoubt of Santoña, if well-prepared, Castro could turn into an almost impregnable fortress, controlling the right side of Napoleon's army.[8]

In the context of the Peninsular War, British presence was visible in Castro as early as August 1808. On 18 August, Major Philip K. Roche, a key figure in Anglo-Cantabrian relations, entered the port while on a mission to support the popular uprising in the Basque Country with arms, ammunition and money.[9] Since then, and until December that year, allied relations with the Cantabrian port can be

---

8   Carmen Gómez Rodrigo, 'Ayuda inglesa a Santander en la Guerra de la Independencia', *XL Aniversario del Centro de Estudios Montañeses* (Santander: CEM, 1976), p.404; Rafael Palacio Ramos, 'Importancia estratégica de Cantabria durante la Guerra de la Independencia: vías de comunicación y plazas fuertes', in Rafael Palacio Ramos (coord.), *Monte Buciero 13: Cantabria durante la Guerra de la Independencia* (Santander: Ayto. de Santoña, 2008), p.240; and, Charles Oman, *A History of the Peninsular War* (Oxford: Clarendon Press, 1914), vol.6, p.265.

9   TNA: WO 1/233, pp.487–504, Roche to Castlereagh, 18 August 1808. Also, Alicia Laspra Rodríguez, *Las relaciones entre la Junta General del Principado de Asturias y el Reino Unido en la Guerra de la Independencia* (Oviedo: Junta General Principado de Asturias, 1999), pp.298–304.

described as occasional, observational, and closely bound to the development of events in the neighbouring Basque provinces. Considering Major General James Leith's correspondence, Castro was an information source about the situation in the Biscay area. Not only that, but French actions there also turned it into an improvised depot to prevent British aid sent to Bilbao from being captured. It was, as well, the most suitable land route from Santander, and was used by Marquis de La Romana's *División del Norte*.[10]

In early 1809, the French triumphal advance across northern Spain meant Castro fell into their hands, which changed the role it played for British officers in the area; thereafter, it became a target for diversionary attacks. Anglo-Spanish amphibious operations were thus constant along the coast in the following months. In 1811, given the relevance Castro was gaining as the key to Santoña from Basque ports and roads, Captain Collier, of the *Surveillante*, insisted in May that, while the fortress was still being fortified by the French, a combined operation should be launched to recapture it.[11] Nonetheless, nothing was done until the spring of 1812, probably coinciding with the 'anti-climax of the allied army in the Peninsula'.[12] The change of circumstances that year due to Napoleon's Russia campaign, as well as the relocation of the main theatre of operations towards inner Spain, exponentially increased the importance of allied actions on the northern coast to divert the *Armée du Nord* from Wellington's advance into Old Castile. The main sea fortresses should then be controlled to prevent French communications and supply deliveries, and to favour the allies' own. Thus, at the end of May, the Admiralty appointed Commodore Sir Home Popham for that mission, which included capturing a seaport to provide the squadron both a safe anchorage and a point to meet the Spanish forces.

Working in co-operation with *Teniente General* Gabriel Mendizábal's Seventh Army, Popham and Francisco de Longa, commander of the Iberian Division, organised an assault on Castro for 6 July. Two days later, after a heavy fire from the squadron and an intense fight on land, the castle was assaulted, and the garrison had no choice but to surrender. The victory, however, did not blind the winners and the French prisoners were treated with all honours because, according to Longa, of the brave defence they had conducted, and immediately embarked for Corunna.[13]

The recapture of Castro was even celebrated by Wellington. On 28 July, he highlighted to Henry Bathurst, Secretary of State for War, the relevance of the action as it had prevented *Général de division* Marie-François A. Caffarelli from assisting the

---

10 TNA: WO 1/229, pp.381–389, Leith to Castlereagh, 18 October 1808; pp.415–416, Leith to Baird, 20 October 1808; pp.451–454, Lefebre to Leith, 25 October 1808; and, pp.459–466, Leith to Castlereagh, 7 November 1808. Also, Alicia Laspra Rodríguez, *La Guerra de la Independencia en los Archivos Británicos del War Office (1808–1809)* (Madrid: Ministerio de Defensa, 2010), p.259; and, Laspra, *Relaciones*, pp.347–351.
11 TNA: WO 1/261, pp.275–302, Collier to Gambier, 30 May 1811.
12 Rory Muir, *Wellington, The Path to Victory 1769-1814* (New Haven: Yale University Press, 2013), p.437.
13 Silvia Gregorio Sainz, 'Sir Home Popham's mission in 1812: Santander a British Logistics Centre?', in Zack White (ed.), *The Sword and the Spirit. Proceedings of the First 'War & Peace in the Age of Napoleon' Conference* (Warwick: Helion & Company, 2021), pp.66–67. Also, Gómez, 'Ayuda', p.406.

Armée du Portugal.[14] The role the Cantabrian town played was not reduced to that. Its control by Spanish forces and the British squadron was crucial, with a threefold objective: first, to hinder French communications and supply deliveries from Bayonne; second, to cut any withdrawal of French soldiers from Santoña; and, mainly, with Santander being in French hands, Castro could function as an allied logistics hub to launch operations in the Basque provinces. This is precisely what happened. This explains why Popham recommended Longa to strengthen its defences. In fact, years later the historian Charles Oman described it as 'the touching-place of British cruisers and the one fortified port which the allies possessed on the Biscay Coast'.[15] However, after consulting the Spanish generals, Popham realised Castro was not the suitable base he was looking for to support Wellington.[16] Unfortunately, Patriot control of the fortress barely lasted 10 months, during which the garrison had to face constant attacks. In fact, from November 1812, and especially in the first half of 1813, the French tried desperately to regain the fortress, proving its strategic importance.

The French Siege: Beyond the Royal Navy's Evacuation Tasks

In December 1812, Napoleon's defeat in Russia struck a heavy blow against the *Grande Armée*. The situation was not better in Spain either; the recent Anglo-Portuguese campaign had not only boosted the Spanish population's morale but also, due to the *Armée du Nord*'s concentration in Burgos to force the raising of the allied siege, the northern provinces were in full insurrection. This hindered French communications with Madrid and so turned Napoleon's attention back to the Peninsula. He decided to reorganise his armies in Spain and gave Caffarelli orders to regain control of the insurgent regions. He was ordered to focus his efforts on retaking the main coastal towns from Santander to San Sebastián before Wellington resumed the campaign in spring. His instructions included replacing the garrison in Santoña and resupplying it.[17] But Caffarelli, busy protecting the convoys from Bayonne, postponed the mission until the beginning of 1813.[18] On 5 January, he finally marched to Santoña but, before entering the fortress, unsuccessfully invested Castro between the 10th and 14th with a seemingly high casualty rate. This attempt was considered relevant enough for Lieutenant Colonel Richard Bourke, British agent in Galicia, to briefly report it on 30 January to Colonel Henry Bunbury, the Under-Secretary of State for War. Even the Spanish Paper *El Diario del Gobierno de Sevilla* echoed it as a morale booster for the Patriot ranks.[19]

---

14   Gurwood (ed.), *Dispatches*, vol.9, pp.317–318, Wellington to Bathurst, 28 July 1812.
15   Oman, *Peninsular War*, vol.6, p.264.
16   Gregorio, 'Sir Home', p.67.
17   Oman, *Peninsular War*, vol.6, pp.252–257; and, Muir, *Wellington*, p.518
18   Rafael Palacio Ramos, *Santoña, plaza napoleónica* (Santoña: Ayuntamiento, 2015), p.145.
19   Maldonado, *Historia*, vol.3, p.365; and, Southey, *History*, vol.3, p.601. TNA: WO 1/267, pp.29–32, Bourke to Bunbury, 30 January 1813; and, Manuel Muriel Hernández and Mariano Cuesta Domingo, 'Noticias sobre Santander y su entorno en la prensa periódica durante la Guerra de la Independencia', *La Guerra de la Independencia y su momento histórico* (Santander: CEM, 1982), vol.1, pp.215–294.

Soon afterwards, considering Caffarelli had failed to quash the Spanish insurgents, Napoleon replaced him with *Général de division* Bertrand Clausel. This time, however, as the emperor insisted on the pressing need to pacify the north and to command every port between the Bidassoa River and Santander (especially, Bermeo and Castro), the *Armée du Portugal* was ordered to assist in the mission.[20] This unconsciously contributed to Wellington's diversionary strategy. Despite these measures, the results were not too different. Closely following and relaying the events in Castro, on 14 April, Bourke announced that, according to the town's Governor, it had been invested on 18 March by 3,000 men led by Palombini. But a week later, already beaten by the garrison and given Mendizábal and Longa's approach, he withdrew towards Bilbao.[21]

Specific dates, although always framed in the second half of March, vary in later historiography. British 'contemporary' printed sources, based on Clausel's engineer Camillo Vaccani's memories, related that on 21 March the *Général en Chef de l'Armée du Nord*, together with Palombini's division and a French battalion, marched to Castro. On arriving there the following day, they determined the siege and assault were impossible with the available means as the fortress had been strongly fortified and the forces of Mendizábal and Campillo were on the watch. News from Bilbao was not promising either since 'El Pastor' was threatening it. Clausel thus postponed the attack and speedily returned to Biscay, leaving in Castro Palombini, who followed him shortly after as the siege was finally raised. Before heading back to Biscay, the latter relieved Santoña and organised with the French Governor the transfer to Castro of a siege train and ammunition.[22] Whether an assault took place or not, the episode shows the importance the Cantabrian town had for both sides, and how the diversionary operations were effective. The ground was getting ready for what was to come in a month.

In these early actions in 1813, no British participation is recorded; and not only that, but also an 'apparent' absence of the Royal Navy vessels off Cantabria was reported. On 7 February, a few days after the French evacuation of Santander, the *Real Consulado* communicated the impossibility of transmitting the news to the allies as no British ships were spotted on the coast.[23] This might be explained, besides the awful weather conditions in winter, by the withdrawal of Popham's squadron from Santander the previous December.[24] Despite this, news of Spanish operations in the northern provinces reached Corunna regularly, as seen in Bourke's correspondence. This suggests that some British ships, although few in number and not always visible, were still occupied with mainly blockading tasks along the coast. The situation would completely change in April; coinciding with Wellington's

---

20 Oman, *Peninsular War*, vol.6, pp.259–260.
21 TNA: WO 1/267, pp.153–156, Bourke to Bunbury, 14 April 1813.
22 Napier, *History*, vol.3, p.300; and, Oman, *Peninsular War*, vol.6, pp.264–266. Spanish sources, however, confirmed the assault, although on different dates, see Maldonado, *Historia*, p.366; Arteche, *Guerra*, pp.68–69; and, more recently, Palacio, 'Importancia', p.240.
23 AHPCan, Real Consulado, 7/27/2, *Real Consulado* in Santander to Cristóbal de Góngora, Spanish Treasury Secretary.
24 Gregorio, 'Sir Home', pp.77–78.

preparations for the forthcoming campaign, orders to intervene were issued at the end of the month.

In early April, British warships were already collaborating with the Spanish forces in the Basque coast to quickly move units, pursued by the enemy, to safer positions. On 9 April, for instance, upon Palombini's imminent attack, several Gipuzkoan battalions were shipped from Lequeitio and Montrico to Castro. So, the Royal Navy's ships were off the coast and participating in combined actions.[25] Nonetheless, the Cantabrian port does not appear in the documents examined in relation to British operations in the Peninsula until the end of the month. On 23 April, planning the new allied offensive into Spain, Wellington urged Collier to prevent enemy communications between Bayonne and Santoña. To do so, Castro was, in Wellington's words, vital and thus Collier was instructed to contact the Spanish garrison there and request their collaboration in the sea blockade. Wellington also warned him that the town had already been attacked and even accurately foresaw a new attempt. The expediency of these directions might also be revealed by the fact that on 6 May, in a dispatch to the Secretary of State for War to suggest that Rear Admiral Thomas Byam Martin be in charge of both the Spanish and Portuguese coasts, Wellington brought them up, probably due to Collier's 'seeming' inactivity.[26] In fact, two days after the loss of Castro, on 13 May, Bourke announced Collier had just sent a sloop with ammunition to collaborate with the Spanish Governor in the town.[27] The chronology of events might thus indicate that the British defence of Castro from 25 April to 11 May did not occur in compliance with those direct orders, but simply responded to the Navy's tasks along the Cantabrian coast.

Wellington was right. From the end of April, French attempts to recover control of Castro intensified; in fact, on 23 April, John Kelly, vice-consul in Gijón, informed the British Consul in Galicia, Asturias and Cantabria that around 10,000 French soldiers had been beaten on their way to invest Castro by Francisco Espoz y Mina.[28] French efforts had failed again, but would soon succeed in May.

Once the requested reinforcements from the *Armée du Portugal* joined Clausel in Vitoria late in April, the siege of Castro was resumed. While Palombini's Italian division would be in charge of besieging the fortress, Foy and Sarrut's men had to cover them and resist any allied disembarkation in the area.[29] Soon after orders were issued, slightly over 11,000 soldiers were on the move. In fact, Captain Robert Bloye, of the *Lyra*, a first-hand witness of the events in the Cantabrian town whose account is central to this essay, informed Collier on 13 May that the enemy had again invested Castro since 25 April and recalled the squadron's collaboration in its defence until 4 May, as he had described in a previous dispatch. This document has not been found yet, but British early actions can be reconstructed thanks to

---

25 Oman, *Peninsular War*, vol.6, p.267.
26 Gurwood (ed.), *Dispatches*, vol.6, p.442, Wellington to Collier, 23 April 1813; and, p.473. Wellington to Bathurst, 6 May 1813. For Wellington's complaints on the 'nominal' sea blockade see Joshua Moon, *Wellington's Two-Front War. The Peninsular Campaigns at Home and Abroad, 1808–1814* (Norman: University of Oklahoma Press, 2011), p.170.
27 TNA: WO 1/267, pp.182–183. Bourke to Bunbury, 13 May 1813.
28 Laspra, *Relaciones*, pp.678–679, Kelly to Allen, 23 April 1813.
29 Napier, *History*, vol.3, p.302.

the campaign diary of *Teniente Coronel del Regimiento de Húsares de Iberia* Pedro P. Álvarez, Governor of Castro, and other 'contemporary' printed sources (mainly, Napier, Oman and Arteche).[30]

On 25 April, with the siege train in Bilbao ready to be shipped, Foy headed for Castro, together with Palombini's and Sarrut's divisions, arriving at its environs on the same day. While Palombini was left to blockade the fortress and ensure the safe arrival of the artillery train, Foy established his headquarters at Cerdigo (probably at Allendelagua); his own division and three Italian battalions would do the siege work. Meanwhile, Sarrut's men were to cover French operations and two Italian units should control the road to Bilbao. Also, on 29 April, Foy ordered the French Governor of Santoña, Charles-Malo-François Lameth, to ship the heavy guns to Islares as soon as he could do so safely. The suitable date would be 4 May.[31]

Besiegers' preparations were thus taking shape by the beginning of May, albeit not without obstacles. The natural position of the citadel on a height was playing against them but, despite being protected by a double wall, reinforced with more than 20 guns, Castro was far from a first-class fortress. It was also covered on land by the Seventh Army.[32] In addition, from 25 April to 4 May, the Spanish garrison (consisting of 1,200 men, according to Bloye)[33] under *Teniente Coronel* Álvarez had not only been preparing the town to first resist an assault and then to slow down the French advance within the walls, but also hindering enemy work by daily sorties and engaging in skirmishes that made the French temporarily retreat. They were not alone in that business; the garrison was supported from the sea by three Spanish *trincaduras* (large two-masted barges), and 'officially' from 3 May, by a British squadron of three brigs – the *Sparrow*, the *Royalist*, and the *Lyra* – commanded by Captains Joseph N. Tayler, James J.G. Bremen and Robert Bloye, by the *Alphea* schooner, Captain McDonald, and even by the privateer brig *Volunteer*.[34] Yet the presence of these ships off the Cantabrian coast and their joint performance with the allies are registered even before that date.

Already on 27 April, as the *Royalist* appeared off Castro, Álvarez requested the captain's assistance in defending the fortress. Only three days later, the brig *Volunteer* also entered the port and was immediately dispatched, together with the Spanish *trincaduras*, on a mission to prevent the French from disembarking the artillery coming from Santoña. The siege train and other means were to proceed by

---

30 Marshall, *Biography*, vol.3, p.140, Bloye to Collier, 13 May 1813. Also, Pedro Álvarez, *Manifiesto que en su defensa y en contextación al que publicó una cabeza exaltada de la villa de Castro Urdiales* (Burgos: Imprenta de Navas, 1813).
31 Napier, *History*, vol.3, p.304; Oman, *Peninsular War*, vol.6, p.271; and, Arteche, *Guerra*, p.69.
32 Napier, *History*, vol.3, p.305; Oman, *Peninsular War*, vol.6, p.265; Arteche, *Guerra*, p.67; and, Palacio, 'Importancia', p.240.
33 Depending on the source, this figure ranges from 1,000 and 1,300. In line with Bloye's report, Southey, *History*, vol.3, p.605, and Napier, *History*, vol.3, p.305 set the number in 1,200. Meanwhile, Oman, *Peninsular War*, p.272, reduced it to 'no more than 1,000 men', probably echoing Arteche, *Guerra*, p.67. On the contrary, Foy's account, in Palacio, 'Importancia', p.240, increased it to 1,300 men. This is also the number stated by Álvarez in *Manifiesto*, p.24, for February.
34 Marshall, *Biography*, vol.3, p.140; and, Southey, *History*, vol.3, p.605. Also, Álvarez, *Manifiesto*, Appendix, pp.19–20 and 22.

sea from Santoña and Bilbao to Islares. However, on 1 May, British warships were blocking the mouth of the River Nervión and, thus, the heavy guns in Bilbao had to be transported on land through almost impracticable paths. Nonetheless, this gave Lameth the opportunity to run his convoy safely across the bay. After that, and coinciding with the anniversary of *Dos de Mayo*, the British squadron was spotted off Castro at dawn, and Álvarez appealed for their assistance. In consequence, the *Lyra*, the *Sparrow* and the *Royalist* officially entered the port on 3 May and a combined action was agreed to attack the enemy on the following day, marking the beginning of British involvement in the defence of the Cantabrian town.[35]

Although Castro had been blockaded from 25 April, the 'proper' siege started on 4 May as the battering train seemed to have finally arrived from Islares. It was therefore possible to start the construction of two batteries placed at key sites: the *Roi de Rome* to the West of the town on the village of Urdiales to flank the fortress' outer wall, the immediate area behind it and Longa's redoubt; on the opposite side, the *Eugène*, to the south-eastern side directed towards the urban area and the castle. The breach battery, the *Impérial*, began to be mounted three days later to the south of the town, taking cover by a house and right in front of San Francisco Convent. Works were not easy, though. On 4 May at dawn, the agreed Anglo-Spanish operation was launched. Once Bloye's squadron took their positions, four Spanish columns sallied out of the town, and an intense fire commenced from the fortress batteries, the castle, and the allied ships. The French responded in similar terms, with the fight reaching a stalemate by the end of the day. The situation was not desperate for the garrison yet, since Álvarez knew reinforcements were on their way.[36]

On 4 May the events in Castro reached a turning point since, according to Bloye, the Spanish garrison did not make any other sortie thereafter due to the enemy's increasing numbers closer to the outer wall. Also, no significant military action seemed to have taken place for several days, with the French mainly concerned at siege work.[37] As a result, as stated by Foy, a trench was opened at around 250 metres from the curtain wall on 6 May, and on the following night, efforts concentrated on erecting the battery to breach it, which took two days to complete.[38] Meanwhile, Bloye discovered the French works on *Roi de Rome* and was determined to hinder their progress. British involvement in the defence of Castro then went a step further: the captain of the *Lyra* ordered the placing of a 24-pounder from the *Sparrow* on the rocky islet of *Los Conejos*, which offered an excellent position to attack the French battery in Urdiales. The next morning, shortly before the allied battery was completed

---

35   Álvarez, *Manifiesto*, Appendix, pp.19–22. For the failed French attempt to ship heavy artillery from Bilbao see Southey, *History*, vol.3, p.605; Napier, *History*, vol.3, p.304; Oman, *Peninsular War*, vol.6, p.271; and, Arteche, *Guerra*, p.69. The date for Lameth's operation in Jacqués V. Belmás, *Journaux de sieges faits ou soutenous par Français dans la Península de 1807 á 1814* (Paris: Didot Frères, 1837), vol.4, p.573.
36   Álvarez, *Manifiesto*, Appendix, pp.22–23; Belmás, *Journaux*, p.575; and, Arteche, *Historia*, pp.72–73. For a map of the French siege of Castro in 1813 see Carmen Delgado Viñas, 'Castro Urdiales (Cantabria), de villa marinera a ciudad de servicios. La transformación urbanística de una ciudad frontera', *Ería*, 86 (2011), p.244.
37   Marshall, *Biography*, vol.3, p.140.
38   Belmás, *Journaux*, p.575.

and the gun positioned, the enemy opened an intense fire against it. Nonetheless, it was successfully returned first by the castle and then by the 24-pounder. Also, at that time, the squadron spotted another French battery being built to the southwest of the town, the *Impérial*, out of reach of the fire from the castle. To counter it, 'a long brass 12-pounder' was also disembarked and mounted on the fort, although it soon burst.[39] Despite this setback, Álvarez praised the effectiveness of the allied battery, which he referred to as *Comodoro Bloye*, as it helped overpower that of the besiegers.[40] Events then took over; on the following days French fire gradually increased its intensity and efficacy.

Early in the morning on 9 May, according to Foy, the batteries *Roi de Rome* and *Eugène* were completed and opened fire, immediately overpowering the Anglo-Spanish artillery. There is a slight disparity among French, Spanish and British reports regarding the chronology of these events, which seems understandable. For example, Álvarez recorded that these French batteries started shooting on 8 May instead and destroyed a large part of the buildings in town. Meanwhile, Bloye marked 10 May as the day the *Eugène* began firing while work on the other batteries continued. Regardless of the date, the fact is that a heavy fire was kept up by both sides and, following the British account, every possible thing was done on their part to collaborate with the Spanish garrison; both to tighten the blockade off Castro to prevent the arrival of the French guns waiting at Portugalete, and to strengthen the fortress defences. To the former end, the *Sparrow* and the *Royalist* were to patrol off the Cantabrian town and, for the latter purpose, it was key to reinforce the allied position on the rocky islet. Thus, Captain Tayler placed a second battery for another 24-pounder and, despite French attempts to harass the works, it was ready to be used on the darkest day for Castro, 11 May. That previous night Álvarez had agreed with the squadron a stratagem to make the enemy believe the defence had been abandoned and the garrison evacuated, but it could not be implemented due to the intense fire from the French batteries at daybreak.[41]

At dawn 11 May, the battery *Impérial* opened fire to breach the town wall. Not being a first-class fortress, it could not resist the attack from 17 siege guns for long and, despite the Anglo-Spanish defence, the wall started to crumble soon after. In fact, by 3:00 p.m., the breach was already 30 feet wide and considered practicable for '20 men abreast'. Therefore, instructions were given to undertake the storm at 8:00 p.m.; two battalions were to assault the breach, and a third one to escalade the wall to the eastward near the *Puerta de Bilbao*. Nothing else could be done by the allies and, their position in *Los Conejos* being untenable, Bloye ordered the withdrawal, putting Captain Tayler in charge of the re-embarkation operation. His directions

---

39 Marshall, *Biography*, vol.3, p.140. Oman, *Peninsular War*, vol.6, p.272, might mislocate the British battery on the islet of Santa Ana placed on the opposite side of the castle.
40 Álvarez, *Manifiesto*, Appendix, p.24.
41 Marshall, *Biography*, vol.3, p.141; Álvarez, *Manifiesto*, Appendix, pp.24–25; and, Belmás, *Journaux*, p.576. According to Álvarez, *Manifiesto*, Appendix, p.24, the British artillery pieces mounted on *Los Conejos* included two 32-pounder carronades, a 24-pounder, and a 6-pounder. However, Bloye in Marshall, *Biography*, vol.3, p.141, only refers to three carronades: two 24-pounders and a 6-pounder. This episode is briefly registered in Napier, *History*, vol.3, p.305, and indirectly in Oman, *Peninsular War*, vol.6, p.272.

included preparing the evacuation of the garrison with Álvarez, which would be effected once all ordnance and the castle had been destroyed. The captain of the *Sparrow*'s role proved key in the defence of Castro, being involved in an epic episode during the withdrawal from the islet. Against all odds, he kept using a carronade, whose chamber had been dented and carriage broken, until the British retreat was completed.[42]

Once the walls were destroyed, resuming Bloye's account of events, French guns were directed towards the town and the castle, heavily bombarding the bridge that connected it with the landing place on the islet of *Santa Ana* to cut off the garrison's withdrawal. The French advance found no obstacle, and the assault started at around 9:00 p.m. The captain of the *Lyra* related that at that time, 3,000 men broke into the town by the breaches as well as by escalading the walls. Despite the garrison's most determined defence, even 'house by house' in Bloye's words, they could not hold their positions longer. In fact, with the assault imminent and aware of the weakness of the outer walls, Álvarez ordered his men to retreat to the second fortification, leaving two companies to defend the town and to cover the withdrawal. Mendizábal's orders were clear: the garrison could not be lost. For three hours, according to the Spanish Governor, these men resisted the charge but, overwhelmed by numbers, had to fall back to the castle. Apparently, therefore, the advance was not unobstructed as Foy claimed. In fact, determined Spanish resistance, and especially that of Álvarez, impressed Bloye, who praised the Governor's performance in his report on the siege.[43] In contrast, some civil representatives – who had already fled to Santander – publicly condemned Álvarez's military decisions during the siege, and even before, due to their terrible results for the town, but that is a different story.[44]

Meanwhile, British boats were ready in *Santa Ana* to receive the garrison onboard, and it was quickly embarked, except for the companies left to cover the Spanish retreat. These men succeeded in throwing every gun left in the castle into the sea but, Bloye regretted, could not blow up the castle before the second fortification was forced. Although half his orders were accomplished in that regard, the evacuation had been a success; the full garrison, including the two companies mentioned, and some inhabitants had been rescued and, afterwards, taken to Bermeo, where they landed on 12 May. The operation was applauded by Álvarez, as well as every action the British squadron had undertaken during the siege, which gives an idea of the real scope and importance of allied participation against the 1813 French siege of Castro.[45]

Nonetheless, the town did not face the same fate and, set on fire in several places, Bloye understood it had been destroyed. He also noted, echoing the intelligence received, that in their advance the French had put every person 'to the bayonet'.

---

42 Belmás, *Journaux*, pp.576–577; Marshall, *Biography*, vol.3, pp.141, 143; and, Álvarez, *Manifiesto*, Appendix, p.25.
43 Marshall, *Biography*, vol.3, p.141; Álvarez, *Manifiesto*, Appendix, pp.25–26 and 34; and, Belmás, *Journaux*, p.578. For the Spanish irruption in the siege see Napier, *History*, vol.3, p.305; and, Arteche, *Guerra*, pp.74–75.
44 Cabarga, *Santander*, p.251. See Álvarez, *Manifiesto*, pp.3–57, for the Governor's defence against the charges.
45 Marshall, *Biography*, vol.3, pp.141–142; and, Álvarez, *Manifiesto*, Appendix, pp.26–27.

Yet, not being there in the aftermath, the captain could not give further details on French cruelty in Castro. Álvarez, in turn, added more details, focusing instead on the horrific noises that filled the air; from women's moans mingled with children's and fathers' cries, to the collapsing of buildings consumed by the intense fire. The fortress had been taken by storm, and those who remained in it underwent the worst horrors. *Capitaine* Nicolas Marcel, who also took part in the siege, recalled how Foy, considering the breach practicable, summoned the garrison three times to evacuate the town, but the Spanish Governor refused to do so and announced it would be defended inch by inch. On the same vein, local historiography describes how the garrison had waved a black flag stating their determination to fight to the death; this has not been found, though, in Álvarez's campaign diary or in his report to Mendizábal.[46] Thus under the 'unwritten' customary laws of war, or as stated in Daly's work the 'law of sack', Castro was a conquered fortress, which meant that besiegers 'had the right to deny quarter to the garrison and sack the town', legitimising at the same time plunder, destruction, murder and rape. As Álvarez's forces had been evacuated, French soldiers turned their fury on the remaining inhabitants. Their fate did not differ much from those siege-related atrocities in Lérida or Tarragona and even, by the allied army, in Ciudad Rodrigo, Badajoz, or San Sebastián.[47]

The harshness of the 18-day siege of Castro can also be understood through casualty figures on all sides, which unsurprisingly vary depending on the source. Bloye reported high losses among the French ranks, although he was unable to make an estimate. This, as he stressed to Admiral Lord Keith on 15 May, had prevented Napoleonic authorities to celebrate 'the conquest'. Captain McDonald of the *Alphea* finally calculated French losses at up to 2,500 men, which Álvarez raised to 3,000. Both figures seem a propagandistic exaggeration, yet the idea of a French 'gloomy' victory spread among the allies. This contrasts with the only 50 soldiers killed and wounded reported by Foy, who at the same time assigned six times more to the Patriot forces. Far from that, Álvarez established the number of Spanish casualties at 100 men. Finally, regarding the British squadron, Bloye communicated 10 wounded marines: '4 in the *Royalist*, and 6 in the *Sparrow*'.[48]

In any case, the official figures reported by all sides seem extremely low considering the constant skirmishes, the intense fire and the fierce resistance for over a

---

46 Marshall, *Biography*, vol.3, pp.141–142; Álvarez, *Manifiesto*, Appendix, pp.25–26 and, 34; Nicolas Marcel, *Campagnes du capitaine Marcel, du 69e de ligne en Espagne et en Portugal (1808–1814), mises en ordre, anotéis at publiées para le commandant Var* (Paris: Libraire Plons, 1813), p.189; and, Cabarga, *Santander*, p.250. The siege, and its terrible results, had an impact on British historiography as it was included, with different degrees of detail, in the most significant printed 'contemporary' sources: Southey, *History*, vol.3, pp.605–606; Napier, *History*, vol.3, pp.304–305; and, Oman, *Peninsular War*, vol.6, pp.272–273.

47 Gavin Daly, *Storm and Sack. British Sieges, Violence and the Laws of War in the Napoleonic Era, 1799–1815* (Cambridge: Cambridge University Press), p.4.

48 Marshall, *Biography*, vol.3, pp.142–143; Álvarez, *Manifiesto, Appendix*, p.27; and, Belmás, *Journaux*, p.578. As for the Spanish, Napier, *History*, vol.3, p.305 increased deaths to 180, which Oman, *Peninsular War*, vol.6, p.273, finds more sensible. The latter author added to British numbers six more injured soldiers and a killed officer and, as for the French, set casualties between 150 and 180.

fortnight. Even more so, in comparison to the alleged 309 civilian deaths as stated in the memorandum the Council in Castro addressed to King Fernando VII in 1819. Undoubtedly, the population took the brunt of the siege, but this figure might have also been slightly inflated in light of Álvarez's reasoning in his *Manifiesto* against civil representatives' accusations. These had already set casualties at about 1,600 people, which the Governor described as a senseless exaggeration since only between 200 and 250 people were estimated to remain at Castro, and from them, the evacuees and survivors should be deducted.[49] Still, a high number compared to civilian deaths resulting from the British storming of Badajoz or San Sebastián.[50]

Once the fall of Castro was a fact, the news spread along the Spanish northern coast and overseas. On 21 May, Kelly communicated it to Consul Allen in Galicia, regretting the atrocities committed by the French forces against the population; and, around a week later, Bourke informed Bunbury about it. On 29 May, Bloye's account of events reached the Admiralty Office in London and was straight away published in the official paper of the Government, *The London Gazette*, and afterwards echoed by most papers in the capital.[51] Surprisingly, already on the move from Freineda (Portugal), Wellington did not acknowledge the loss of Castro until 31 May, as he had heard only rumours, which were confirmed on 6 June in Palencia when he received the official report.[52] Thus, the British squadron was fully engaged in the defence of Castro, not just in the successful evacuation of soldiers and civilians. But their role did not end on 11 May, as Royal Navy ships were crucial for a later episode: the recapture of the fortress.

British Prompt Takeover: Uncovering the Atrocities

Having taken Castro, on 12 May 1813 Foy garrisoned it with the Italian division, which was tasked with watching the coast and keeping communication with Santoña open. He then resumed operations to restrain the northern insurrection, leaving the place in pursuit of the Spanish forces that had jeopardised the siege. Together with Sarrut, Foy forced them to retreat to Santander and, soon after, headed to Bilbao.[53] Nonetheless, a major threat was advancing unnoticed through Spain; Wellington had launched the new campaign on 13 May. The Anglo-Spanish actions in Castro had contributed to it by diverting the attention of French forces from key preparatory movements for over a fortnight. In turn, the allies' sudden incursion in the Basque

---

49 Álvarez, *Manifiesto*, pp.49–50. For the 1819 memorandum see 'Castro Urdiales durante la Guerra de la Independencia', <www.1808-1814.org/articulos/castrourdiales.html>, accessed 12 December 2012.
50 Muir, *Wellington*, p.456; and, Iñaki Egaña, *Donostia 1813* (Donostia: Txertoa, 2012), pp.212–213.
51 Laspra, *Relaciones*, pp.680–681, Kelly to Allen, 21 May 1813; and, TNA: WO 1/267, pp.201–204, Bourke to Bunbury, 29 May 1813. Also, *The London Gazette* (hereafter *LG*), n.16733, 25–29 May 1813, pp.1013–1015.
52 Gurwood (ed.), *Dispatches*, vol.10, pp.408–410 and 421–423, Wellington to Bathurst, 31 May and 6 June 1813.
53 Belmás, *Journaux*, pp.566–567; Napier, *History*, vol.3, p.305; and, Oman, *Peninsular War*, vol.6, p.274.

country in June favoured the return of the Cantabrian town to Patriot hands; quite a disregarded episode, particularly with regards to British involvement.[54]

At that time the Royal Navy's ships remained vigilant off the northern coast and continued their collaboration with the Spanish regulars eastward. In fact, on 10 June, Captain Tayler of the *Sparrow*, responding to an urgent request for assistance, embarked Miguel Artola's battalion in Lequeitio and shipped it to Santander. On his return and sailing off Castro on 22 June, he noticed the French garrison was preparing to abandon the fortress. The tight blockade it had been subjected to by the squadron at sea and on land by Mendizábal's forces had finally yielded results.[55] The allies' success in Vitoria just one day before, 21 June, no doubt aggravated the garrison's anxiety since Wellington's progress forward would mean their land routes to France were blocked and, if the fortress were invested without the suitable means to hold its defence, they would be lost. Retreating to Santoña, the strongest French-held redoubt on the coast, seemed the most sensible solution.

Anticipating the French commander's next steps, Tayler feinted an attack towards Castro, forcing the garrison to retreat hastily without destroying the artillery and the castle. He then occupied and garrisoned the fortress, hoisting the Spanish standard. On 23 June, the captain of the *Sparrow* communicated the positive news to the Governor of Bilbao and committed himself to protecting the place until a Spanish force took over. Mendizábal's men finally did so on 25 June, at which point Tayler headed to San Sebastián. For two days Castro had been under British control, but no fears of it being turned into a new Gibraltar have been found, unlike what would happen in Santoña in 1814. Thus, thanks to British actions a strategic enclave to launch the sieges of San Sebastián and Santoña was recovered. Also, it helped support Wellington's advance towards the Pyrenees as the port of Santander was turning into a British logistics base on the coast. This was not a unique event though, as Collier informed Keith on 25 June that the French had abandoned the main fortresses from Guetaria to Santoña, which gives an idea of the French forces' weakness at the time.[56]

Based at Castro for a few days, Tayler witnessed the destruction of the town and the atrocities the French troops had committed against civilians, which he portrayed as unbelievable to the Governor of Bilbao. He then described the dreadful scene: more than 3,000 people, including babies, had been mercilessly murdered; women and children, regardless of their age, raped; and, the few survivors thrown into misery as the town had been burnt down.[57] The figure of victims seems exaggerated

---

54   Among the printed sources examined, both British and Spanish, only Southey, *History*, vol.3, p.639, referred to the British takeover of Castro in June. Napier, *History*, vol.3, p.348, and Maldonado, *Historia*, p.394, just mentioned the French evacuation without offering more details on it.
55   Marshall, *Biography*, vol.3, pp.143–144; *LG*, n.16746, 3 July 1813, p.1270, Collier to Keith, 25 June 1813.
56   Marshall, *Biography*, vol.3, p.144, Tayler to the Governor of Bilbao, 23 July 1813; *LG*, n.16746, 3 July 1813, p.1270, Collier to Keith, 25 June 1813; and, Gurwood, *Dispatches*, vol.10, pp.461–462, Wellington to Collier, 26 June 1813; and, vol.11, pp.196–198, Wellington to Aylmer and Kennedy, 16 October 1813.
57   Marshall, *Biography*, vol.3, p.144.

considering the number of civilians in the town, but the image is strongly reminiscent of other stormed Spanish towns. Similarly, off Castro on 25 June, Collier also saw first-hand instances of French cruel behaviour there, even having the opportunity to interview a few surviving women. Although he refused to relate in writing the atrocities as being 'too shocking', he revealed some important data: first, 'five-sixths' of the town was destroyed; and, also, the 'carnage' continued after the storm. His choice of words reveals the impression the crimes left on him. Yet these actions seemed not to have gone unpunished; 14 of the 'savage authors' were apprehended in Bilbao and 'deservedly put to death'.[58]

News of the French evacuation of Castro, and the besiegers' crimes, again travelled fast. Collier contributed to it when meeting Sir August S. Frazer on 29 June on his way from Santander to Deba (Gipuzkoa). They shared the latest updates on the issue and, interestingly, Collier explained such cruelty as the population had not warned the enemy the bridge between the castle and *Santa Ana* had been destroyed. Thus, as Frazer recalled in a private dispatch the following day, some French soldiers had died falling down a ravine while charging forward.[59] Kelly gave a similar revenge motivation on 21 June. According to information dated 16 May in Santander, he attributed that brutality, which is described as 'usual', to the garrison's evacuation in British ships.[60] Both reasons might have infuriated stormers, but extreme violence seems to have been deep in their spirits before the assault, as the experience of sieges during the Peninsular War suggests. Yet, according to Oman, Álvarez's men could be evacuated more easily because the French had delayed the advance to run wild, committing crimes comparable to those of Badajoz, but 'on a small scale'.[61]

Whether these atrocious acts were exaggerated is difficult to know. That was how the scene was sketched by witnesses from both sides and, also, the way it was portrayed in the press and in subsequent printed works. The French crimes in Castro thus turned into an interesting issue to be covered by the papers in London as it helped illustrate the savage 'other', but it was quickly overshadowed by far more positive news: Wellington's triumph in Vitoria. In fact, the news regarding the Cantabrian town arrived at the Admiralty Office on 3 July, and was immediately included in an extraordinary issue of *The London Gazette*. That same day, Wellington acknowledged the French evacuation of Castro in a dispatch to Bathurst from Ostiz (Navarre) that was also published in the Government's paper later that month.[62]

---

58   *LG*, n.16746, 3 July 1813, p.1270, Collier to Keith, 25 June 1813.
59   Edward Sabine, *Letters of Colonel Sir Augustus Simon Frazer, K.C.B.* (London: Longman, Brown, Green & Roberts, 1859), p.177.
60   Laspra, *Relaciones*, pp.680–681, Kelly to Allen, 21 May 1813.
61   Oman, *Peninsular War*, vol.6, p.273.
62   *LG*, n.16746, 3 July 1813, p.1270, Collier to Keith, 25 June 1813. Also, Gurwood, *Dispatches*, vol.10, pp.501–507, Wellington to Bathurst, 3 July 1813.

Impact on the London Papers: Defence, Evacuation and Barbarity

Between January and July 1813, the papers published in London relayed information on the development of events in Castro, not only reproducing extracts from *The London Gazette*, but also including particulars received through other channels (Spanish and French papers and mails, official bulletins, etc.). In fact, an analysis of the *British Newspaper Archive* website's catalogue shows that in that time frame, there are around 196 references (concentrated in May and July) to 'Castro' and 'Castro Urdiales' in 20 papers from different ideologies and scopes.[63] Although not comparable to the numerous allusions to Vitoria or San Sebastián in the same period, these figures are significant since they show the impact operations there had in Great Britain. Information regarding Castro in 1813 included the French attempts to regain the fortress in January and March, the Anglo-Spanish defence and the inevitable final evacuation in May, as well as the atrocities committed there by French troops, and finally, the British retaking of the town in June.

On 25 January the placename Castro appears for the first time in 1813 in the tory paper *The Sun*, although in passing, in connection with Caffarelli's mission to relieve and resupply the garrison in Santoña. As part of that manoeuvre, already on 11 February, *The Pilot* published that a French force had unsuccessfully invested the fortress early that month, as described in a dispatch dated 30 January in Corunna.[64] The Cantabrian town is referred to next in the Government's official paper, *The London Gazette*, on 27 April in the context of Palombini's actions in the area. On its front page, it included the report, dated 14 April, in which Bourke reported a new failed attack against Castro on 18 March. This document interestingly denounced the 'excesses and barbarities' committed by French forces, specifically 'Italian troops', in the neighbouring villages during the siege.[65] Early operations in the town, as another example of Spanish resistance, were considered relevant enough to be transmitted to readers, although no British participation was recorded by then. In fact, the same dispatch was reproduced, without adding evaluation comments, in eight other papers published in London from the conservative *The Sun* to even a representative paper of a Spanish liberal exile, José M. Blanco White's *El Español*.[66]

The news soon changed tone as, according to the liberal *Morning Chronicle* and the *Saint James's Chronicle* on 8 May, the French wanted to turn Castro into 'another

---

63 BNA: *Anti-Gallican Monitor, Bell's Weekly Messenger, Champion, Cobbet's Weekly Political Register, Englishman, The Examiner, Johnson's Sunday Monitor, London Chronicle, London Courier and Evening Gazette, London Moderator and National Adviser, Morning Chronicle, Morning Post, National Register, The News, Pilot, Public Ledger and Daily Advertiser, Saint James's Chronicle, Star, Statesman*, and *Sun*. Research has been hindered by the different English spellings of the Spanish placename 'Castro Urdiales'.
64 *Sun*, 25 January 1813, p.2. Also, *Statesman* and *Pilot*, 25 January 1813, p.2; and, *Saint James's Chronicle*, 26 January 1813, p.2. And, *Pilot* and *Saint James's Chronicle*, 11 February 1813, pp.3 and 4, respectively.
65 *LG*, n.16723, 24–27 April 1813, p.813.
66 *Sun, Star, Pilot, Saint James's Chronicle, The News, Anti-Gallican, Englishman, National Register, Johnson's Sunday Monitor*, and *Bell's Weekly Messenger*, 28 April 1813, pp.4, 2, 1, 1, 3, 6, 4, 6, 2 and 6. Also, José M. Blanco White, *El Español. Abril de 1813* (London: R. Juigné, 1813), p.321.

Pancorbo or Castle of Burgos', recalling therefore Wellington's failure to capture the Castilian castle at the end of 1812.[67] In the following weeks, short notes – with confusing details – on the town's fate filled periodicals echoing the intelligence received by the Corunna and Cádiz mail. Thanks to HM Packet *Lady Mary Pelham*, it was first known that French troops were moving towards Castro and, then, that the siege had begun on 28 and 29 April.[68] At the same time, though, the news of the fall of the Cantabrian town into French hands started to leak in the papers. In fact, on 20 May, *The Sun* stated that, according to private accounts from Corunna, the French had entered Castro with a high cost in casualties. And, drawing on the Spanish papers, the *Saint James's Chronicle*, among other periodicals, also relayed the loss of the fortress.[69] No trace of British involvement in the defence is discernible until 28 May, when 'several British brigantines' were said to be off that port to succour the garrison.[70] Curiously enough, similar information about the presence of allied ships in Castro was published that day in the Spanish liberal newspaper *El Conciso*.[71]

The French recapture of Castro was finally confirmed to the reading public by *The London Gazette* published on 29 May. The issue included the account written by Bloye, and received that day in the Admiralty Office, detailing the development of events in the town while highlighting the Navy's participation in its defence and evacuation.[72] Unsurprisingly, 17 London periodicals immediately followed and fully reproduced the report, which was introduced in several cases with an interesting wording. For example, on 30 May, the *National Register*, under a heading entitled 'Fall of Castro', described the Spanish garrison's resistance as 'the most gallant', pointing out British participation. Similarly, Lewis Goldsmith's *Anti-Gallican Monitor* depicted Álvarez's men as 'brave' who 'assisted by the British seamen, most obstinately maintained a very unequal contest'. Yet, the most remarkable part of this introduction is the way the editor placed the loss of Castro on the same level as the episodes of Zaragoza and Gerona in the collective memory of the Peninsular War. On that same day, *Johnson's Sunday Monitor* published the official account as well, emphasising the 'most gallant defence' of the Spanish in collaboration with Bloye's squadron, and referring briefly to French cruel behaviour against civilians as follows: 'and the people put indiscriminately to death by the desperate conquerors!'. It is surprising, though, that no other paper mentioned this information in their initial summaries of the *Gazette*'s contents.[73] In these periodicals, British perfor-

---

67 *Morning Chronicle* and *Saint James's Chronicle*, 8 May 1813, pp.2 and 4.
68 *Star, Pilot* and *Statesman*, 19 May 1813, p.3; *Morning Chronicle, Morning Post* and *Public Ledger*, 20 May 1813, pp.2, 2 and 3; and, *Englishman*, 23 May 1813, p.2.
69 *Sun*, 20 May 1813, p.2. Also, in *Saint James's Chronicle*, 20 May 1813, p.4; and, *Public Ledger*, 21 May 1813, p.2.
70 *Sun, Pilot* and *Star*, 28 May 1813, pp.2, 3 and 3; and, *Morning Post*, 29 May 1813, p.2.
71 *El Conciso* (Cádiz), 28 May 1813, p.5.
72 LG, n.16733, 25–29 May 1813, pp.1013–1015.
73 *National Register, Anti-Gallican Monitor*, and *Johnson's Sunday Monitor*, 30 May 1813, pp.3, 5 and 3, respectively. Also, *Englishman, Champion, The News*, 30 May 1813, pp.3, 6, and 4; *Morning Chronicle, Statesman, Star, Pilot, Morning Post, Sun, London Courier* and *Public Ledger*, 31 May 1813, pp.2, 4, 1–3, 2 and 4, 2–3, 4, 1–3, and 2; *Saint James's Chronicle*, 1 June 1813, p.1; and, *London Moderator*, 2 June 1813, p.4.

mance is praised in quite similar terms. In fact, on 31 May, the *Statesman* and *The Pilot* attributed the successful evacuation of the garrison and the population to the 'active and persevering efforts' of the three Royal Navy ships.[74] Newspaper readers in London were thus aware of the squadron's full involvement in the defence of Castro. By contrast, Spanish papers limited allied actions to the evacuation tasks. On 11 June, *El Conciso* communicated the fall of the town, focusing on its destruction and the murder of up to 1,000 people, but British participation was not revealed until the issue published on the following day. No other allusion to that has been found so far in the papers published in Spain.[75]

'Gallant' and 'brave' seem to be the most repeated adjectives in London periodicals to describe Spanish resistance, as well as British performance, but it might not reflect the editors' actual reaction to the episode. An analysis shows that they were just mirroring the qualifiers in the official report. Nonetheless, a genuine – although moderate – response was appreciated in the *Anti-Gallican Monitor* and in the *Johnson's Sunday Monitor*'s comments mentioned before. The 'limited attention' might be explained considering the wider context of the Peninsular War; the positive news regarding Wellington's unobstructed advance through Spain was reaching Great Britain around the same time and thus drawing the newspapers' attention, and retaining in turn that of the reading public.

Further details on the fall of Castro were still emerging in the London papers after the end of May. It appeared, for example, in three different snippets in June. First, on the 10th, the *Morning Post* referred to Foy's 'uninteresting' report on the taking of the fortress published in the *Gazette de France*. With the public already informed on the events, no British periodical apparently included it.[76] Five days later, the *Statesman* confirmed, as received by the Lisbon and Corunna Papers, the 'horrors of Castro Urdiales', but without going into detail.[77] Finally, on 30 June, *The Sun* named the Cantabrian town in an official bulletin relating Wellington's successful operations in Burgos and his advance towards the Ebro, which the editor described as the 'Most Glorious News'. In such a positive context, it is mentioned to emphasise the 'severe' casualties suffered there by the French, which prevented it from being considered a victory.[78] Interestingly, a connection is established between diversion operations in Cantabria and Wellington's successful advance.

The following news about Castro relayed its evacuation by the French, after the French defeat in Vitoria. Before any official account was received, *The Pilot* reported on 3 July that the garrison in the fortress, according to the *Lapwing* packet, had started a retreat due to Wellington's presence in Vitoria on 20 May, and upon the general withdrawal of the French.[79] That night, 3 July, an extraordinary issue of *The*

---

74 *Statesman*, *Pilot* and *Morning Post*, 31 May 1813, pp.4, 2, and 3.
75 *El Conciso* (Cádiz), 11 and 12 June 1813, pp.7 and 6, respectively.
76 *Morning Post,* 10 June 1813, p.3. Also, *Pilot, Star, Saint James' Chronicle* and *Statesman*, 10 June 1813, pp.2, 3, 4, and 1, respectively; and, *National Register*, 13 June 1813, p.3.
77 *Statesman*, 15 June 1813, p.4.
78 *Sun*, 30 June 1813, p.2. Also, *Star* and *London Courier*, 30 June 1813, pp.3, and 4; *Morning Post, Pilot, Saint James's Chronicle* and *Statesman*, 1 July 1813, pp.3, 2, 4, 2, respectively; and, *The News*, 4 July 1813, p.5.
79 *Pilot* and *London Courier*, 3 July 1813, pp.2, and 3.

London Gazette was published to communicate the allied victory in the capital city of Álava. The paper also contained a copy of a dispatch from Collier reporting the enemy evacuation of Castro on 22 June and, more importantly, the retaking of the fortress by Tayler. In it, the Italian division's withdrawal to Santoña was presented as an exclusive result of the effective British sea blockade, perhaps trying to counter the criticism about Collier's handling of the naval control of the northern coast. In addition, the *Gazette* offered a few more details of French behaviour during and after the storming of Castro: first, the town had been reduced to rubble, just slightly over 15 percent remained standing; and the population suffered such crimes that could not be reproduced.[80]

The next morning, 4 July, five 'minor' papers printed in the British capital immediately copied the information in *The London Gazette* regarding the French evacuation of Castro, as well as the particulars on the issue published in the Portuguese and Spanish papers received in England.[81] More popular periodicals, such as the *Morning Chronicle* or *The Sun*, among others, waited to relay the latest updates until the following day. Here, too, no evaluative comments were added, as the focus was on the allied success in Vitoria. In that respect, *The Sun* should be pointed out since it included some genuine details, although again barely evaluative, about the population in Castro. The 'enormities' were also described as 'too shocking to relate', but the article centred on the 'few surviving inhabitants' left to live in misery, whom the squadron assisted by sending a surgeon and distributing bread until provisions from Bilbao arrived.[82]

No more information on those crimes appeared in the press thereafter. There is just an interesting remark included in the *Morning Post* on 23 July that might have helped its readers comprehend the extent of the horrors suffered in Castro. As described in a dispatch from an officer of the *Surveillante* dated off San Sebastián 8 July, the French had hidden, upon their retreat from Guetaria, a match burning in the magazine behind two barrels of wine to attract the inhabitants. When it exploded, many were killed. However, significantly, it added that this episode was 'nothing to what they did at Castro'.[83] Thus, at least among the officers commissioned on the Spanish northern coast, the atrocities in the Cantabrian town were part of that collective imagination of French brutality. How far this entered the public consciousness is more difficult to determine, as no literary creations (for example, occasional poetry) on the issue have been found.

On the contrary, the French withdrawal from Castro is found once more in the London periodicals, but merely as part of allied operations after Vitoria. On 19 July, an official bulletin from the War Department was published in *The Pilot* and in the *London Courier*, including Wellington's account, dated 3 July and received in the

---

80 *LG*, n.16746, 3 July 1813, p.1270.
81 *Anti-Gallican Monitor, Bell's Weekly Messenger, Johnson's Sunday Monitor, Englishman* and *National Register*, 4 July 1813, pp.6, 7, 3, 4 and 7.
82 *Sun*, 5 July 1813, p.3. Also, *Morning Chronicle, Star, London Courier* and *Statesman*, 5 July 1813, pp.2, 4, 2 and 3; *Saint James's Chronicle*, 6 July 1813, p.2; *London Moderator*, 7 July 1813, p.8; and, *Champion*, 11 July 1813, p.3.
83 *Morning Post*, 23 July 1813, pp.2–3. Also, *London Courier* and *The Pilot*, 23 July 1813, pp.2, and 3; and, *Saint James's Chronicle*, 24 July 1813, p.3.

fourth week of July in London.[84] Although the retreat to Santoña was registered plainly, its inclusion is fairly significant because Wellington acknowledged the fact at that time, even if the news had arrived at the Admiralty Office on 25 June. And it also helps fit operations in Castro into the broader campaign in 1813.[85] On the following day, 20 July, *The London Gazette* inserted the dispatch and, immediately afterwards, the papers in the British capital echoed it.[86]

In Great Britain, the events in Castro were thus considered relevant enough to be included in one of the most well-known yearly publications in London, *The Annual Register*. In the volume for 1813, the three key moments in the Cantabrian town (the French early siege attempts, the assault and its occupation, and finally the evacuation) were outlined, without value judgements, based on the official dispatches published in *The London Gazette*.[87] However, the information is so simplified that neither the Anglo-Spanish defence of the town nor French crimes there were mentioned. This suggests that its importance was relative, being limited to a background action in the context of Wellington's new campaign.

Conclusions

This chapter has revisited the British and Spanish accounts of the French siege of Castro in May 1813. The focus has been on two key first-hand documents, both of which are rather forgotten: Captain Bloye's report and the campaign diary of the Spanish Governor of Castro. All allusions to the town in the official reports and correspondence by Wellington and the officers stationed on the Spanish northern coast have been analysed, as well as the references to the episode in the London papers and in the main 'contemporary' printed sources on the Peninsular War. These documents have helped determine the actual role the allies had in the defence of the besieged fortress, which is usually oversimplified, and also the impact the event had on Anglo-Spanish-Portuguese operations.

From 1808, and especially from 1812, the British presence in Castro can be traced. Due to its natural characteristics, its central position on the coast and the growing importance of Santoña, it proved key in order to oversee and support Spanish operations in the Basque Country, and Popham therefore participated in its recapture. Actions in the area proved essential to divert French forces from Anglo-Portuguese movements over an extended time, especially in the spring of 1813. During the early months of that year, British ships not only fulfilled blockading tasks but also co-operated with the Spanish forces. Their relations with the town were not limited to the

---

84 *Pilot* and *London Courier*, 19 July 1813, p.3. Also, *Morning Post* and *Morning Chronicle*, 20 July 1813, pp.4, and 2; *Johnson's Sunday Monitor, National Register, Englishman, The News, Bell's Weekly Messenger, The Examiner* and *Anti-Gallican Monitor*, 25 July 1813, pp.4, 5, 4, 3, 2, 7, and 4. Also, Blanco, *El Español. Julio 1813*, p.61.
85 Gurwood (ed.), *Dispatches*, vol.6, p.578, Wellington to Bathurst, 3 July 1813.
86 *LG*, n.16753, 20 July 1813, p.1402. Also, *Star, Sun, Saint James's Chronicle* and *Statesman*, 20 July 1813, pp.1, 2, 3, and 2; *Morning Post* and *Morning Chronicle*, 21 July 1813, p.2.
87 Anon., *The Annual Register or a view of the History, Politics and Literature for the Year 1813* (London: Baldwin, Cradock and Joy, 1814), pp.140, 142, 144–145 and 176.

events of 11 May or to evacuation tasks. The *Lyra*, *Sparrow* and *Royalist* were fully involved in the town's defence from the first day of the French siege. The squadron blocked the heavy guns in Bilbao, attacked enemy works and countered their fire from the sea and from the *Comodoro Bloye* battery, reinforced the walls with their own artillery, and, when nothing else could be done, guaranteed the evacuation of soldiers and civilians. That was not all; in June, the captain of the *Sparrow* hastened the French retreat and garrisoned the fortress for a couple of days, recovering an essential enclave. The Governor of Castro praised British performance over this time, but Spanish periodicals barely echoed it.

Newspaper readers in London, on the contrary, were fully informed of the development of events in the remote Cantabrian town, and certainly of British involvement, as a wide range of papers in the capital, regardless their political ideology (although leaning towards the conservative side), considered them relevant enough to be published. However, the news soon coincided with Wellington's triumph in Vitoria, which might explain the scarce reactions by the editors and the readers to the information, preventing us from determining the real impact on public opinion. As these reports were mainly echoing *The London Gazette*, when considered with articles on the allies' successful advance, it can be seen that there was an underlying interest to publish any promising news on the conflict, after wearying years of war, to bolster public support for British intervention in Spain.

Nonetheless, the combined efforts to defend the fortress show the importance of Castro, being essential to both support the new function Santander was about to serve and the forthcoming sieges of San Sebastián and Santoña. The role it played in the broader context of the Peninsular War is also noteworthy: military operations there had diverted three important French divisions for over a fortnight from Longa's actions in Álava and especially from Wellington's preparations to advance towards the Duero River. They were then ready to strike a heavy blow to the emperor. The analysis of British sources has also contributed to completing the gaps in the history of the conflict in Cantabria and to better understand the implications of events there for the wider national context. In a broader sense, the episode in Castro in 1813 also provides more data on Anglo-Spanish relations during the conflict, which still need to be researched further.

Bibliography

Manuscripts
The National Archives of the United Kingdom (TNA)
    WO 1/233; 1/229, 1/261; 1/267
Archivo Histórico Provincial de Cantabria (AHPCan)
    Real Consulado, 7/27/2

Primary Source Books
Álvarez, Pedro P., *Manifiesto que en su defensa y en contextación al que publicó una cabeza exaltada de la villa de Castro Urdiales* (Burgos: Imprenta de Navas, 1813)

Anon., *The Annual Register or a view of the History, Politics and Literature for the Year 1813* (London: Baldwin, Cradock and Joy, 1814)

Blanco White, José M., *El Español. Abril de 1813* (London: R. Juigné, 1813)

Gurwood (ed.), John, *The Dispatches of Field Marshall the Duke of Wellington K.G. during his various Campaigns in India, Denmark, Portugal, Spain, the Low Countries, and France. From 1799 to 1818* (London: J. Murray, 1852)

Laspra Rodríguez, Alicia, *Las relaciones entre la Junta General del Principado de Asturias y el Reino Unido en la Guerra de la Independencia* (Oviedo, Junta General del Principado de Asturias, 1999)

Laspra Rodríguez, Alicia, *La Guerra de la Independencia en los Archivos Británicos del War Office (1808-1809)* (Madrid: Ministerio de Defensa, 2010)

Marcel, Nicolas, *Campagnes du capitaine Marcel, du 69e de ligne en Espagne et en Portugal (1808-1814), mises en ordre, anotéis at publiées para le commandant Var* (Paris: Libraire Plons, 1813)

Marshall, John, *Royal Naval Biography; or, Memoirs of the Services of all the flag-officers, superannuated rear-admirals, retired-captains, post-captains, and commanders, whose names appeared on the Admiralty List of Sea-Officers at the commencement of year 1813, or who have since been promoted* (London: Longman, 1829)

Sabine, Edward, *Letters of Colonel Sir Augustus Simon Frazer, K.C.B., commanding the Royal Horse Artillery in the Army under the Duke of Wellington written during the Peninsular Wars and Waterloo Campaigns* (London: Longman, Brown, Green & Roberts, 1859)

Secondary Sources

Belmás, Jacqués V., *Journaux de sieges faits ou soutenous par Français dans la Península de 1807 á 1814* (Paris: Didot Frères, 1837)

Gavin, Daly, *Storm and Sack. British Sieges, Violence and the Laws of War in the Napoleonic Era, 1799-1815* (Cambridge: Cambridge University Press)

Delgado Viñas, Carmen, 'Castro Urdiales (Cantabria), de villa marinera a ciudad de servicios. La transformación urbanística de una ciudad frontera', *Ería*, 86 (2011), pp.237–270

Egaña, Iñaki, *Donostia 1813* (Donostia: Txertoa, 2012)

Gómez de Arteche, José, *Guerra de la Independencia. Historia Militar de España de 1808 a 1814* (Madrid: Depósito de la Guerra, 1902)

Gómez Rodrigo, Carmen, 'Ayuda inglesa a Santander en la Guerra de la Independencia', *XL Aniversario del Centro de Estudios Montañeses* (Santander: CEM, 1976)

Gregorio Sainz, Silvia, 'Sir Home Popham's mission in 1812: Santander a British Logistics Centre?', in Zack White (ed.), *The Sword and the Spirit. Proceedings of the First 'War & Peace in the Age of Napoleon' Conference* (Warwick: Helion & Company, 2021)

Greenblatt, Stephen, and Catherine Gallagher, *Practicing New Historicism* (Chicago: University of Chicago Press, 2000)

Moon, Joshua, *Wellington's Two-Front War. The Peninsular Campaigns at Home and Abroad, 1808-1814* (Norman: University of Oklahoma Press, 2011)

Muir, Rory, *Wellington, The Path to Victory 1769-1814* (New Haven: Yale University Press, 2013)
Muriel Hernández, Manuel, and Mariano Cuesta Domingo, 'Noticias sobre Santander y su entorno en la prensa periódica durante la Guerra de la Independencia', *La Guerra de la Independencia y su momento histórico* (Santander: CEM, 1982)
Muñoz Maldonado, José, *Historia política y militar de la Guerra de la Independencia de España contra Napoleón Bonaparte desde 1808 a 1814* (Madrid: José Palacio Ramos, 1833)
Napier, W.F.P., *History of the War in the Peninsula and in the South of France, from the year 1807 to the year 1814* (Brussels: Meline, 1839)
Palacio Ramos, Rafael, 'Importancia estratégica de Cantabria durante la Guerra de la Independencia: vías de comunicación y plazas fuertes', in Rafael Palacio Ramos (coord.), *Monte Buciero 13: Cantabria durante la Guerra de la Independencia* (Santander: Ayuntamiento de Santoña, 2008)
Palacio Ramos, Rafael, *Santoña, plaza napoleónica* (Santoña: Ayuntamiento, 2015)
Simón Cabarga, José, *Santander en la Guerra de la Independencia* (Santander: José Simón Cabarga, 1968)
Southey, Robert, *History of the Peninsular War* (London: John Murray, 1832)
Oman, Charles, *A History of the Peninsular War* (Oxford: Clarendon Press, 1914)

From Reason to Revolution – Warfare 1721-1815

http://www.helion.co.uk/series/from-reason-to-revolution-1721-1815.php

The 'From Reason to Revolution' series covers the period of military history 1721–1815, an era in which fortress-based strategy and linear battles gave way to the nation-in-arms and the beginnings of total war.

This era saw the evolution and growth of light troops of all arms, and of increasingly flexible command systems to cope with the growing armies fielded by nations able to mobilise far greater proportions of their manpower than ever before. Many of these developments were fired by the great political upheavals of the era, with revolutions in America and France bringing about social change which in turn fed back into the military sphere as whole nations readied themselves for war. Only in the closing years of the period, as the reactionary powers began to regain the upper hand, did a military synthesis of the best of the old and the new become possible.

The series examines the military and naval history of the period in a greater degree of detail than has hitherto been attempted, and has a very wide brief, with the intention of covering all aspects from the battles, campaigns, logistics, and tactics, to the personalities, armies, uniforms, and equipment.

Submissions
The publishers would be pleased to receive submissions for this series. Please email reasontorevolution@helion.co.uk, or write to Helion & Company Limited, Unit 8 Amherst Business Centre, Budbrooke Road, Warwick, CV34 5WE

You may also be interested in:

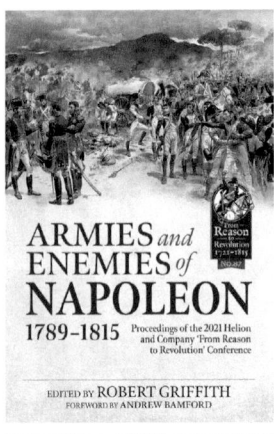
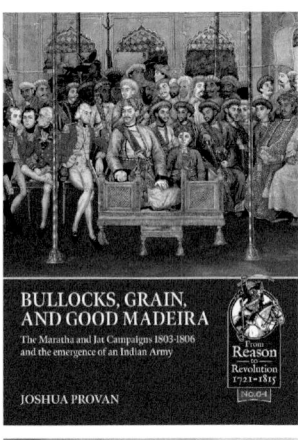
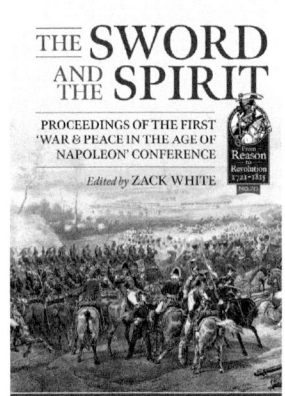
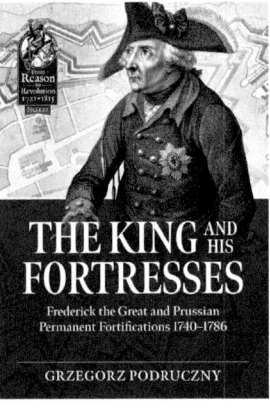
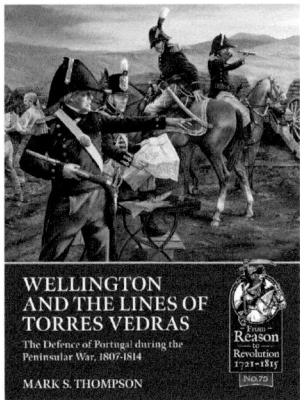
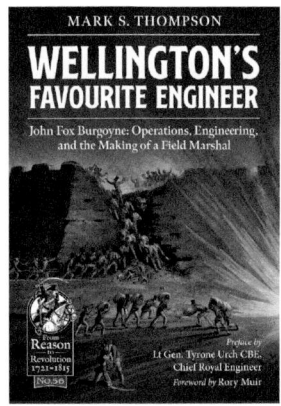